Springer Theses

Recognizing Outstanding Ph.D. Research

Aims and Scope

The series "Springer Theses" brings together a selection of the very best Ph.D. theses from around the world and across the physical sciences. Nominated and endorsed by two recognized specialists, each published volume has been selected for its scientific excellence and the high impact of its contents for the pertinent field of research. For greater accessibility to non-specialists, the published versions include an extended introduction, as well as a foreword by the student's supervisor explaining the special relevance of the work for the field. As a whole, the series will provide a valuable resource both for newcomers to the research fields described, and for other scientists seeking detailed background information on special questions. Finally, it provides an accredited documentation of the valuable contributions made by today's younger generation of scientists.

Theses are accepted into the series by invited nomination only and must fulfill all of the following criteria

- They must be written in good English.
- The topic should fall within the confines of Chemistry, Physics, Earth Sciences, Engineering and related interdisciplinary fields such as Materials, Nanoscience, Chemical Engineering, Complex Systems and Biophysics.
- The work reported in the thesis must represent a significant scientific advance.
- If the thesis includes previously published material, permission to reproduce this must be gained from the respective copyright holder.
- They must have been examined and passed during the 12 months prior to nomination.
- Each thesis should include a foreword by the supervisor outlining the significance of its content.
- The theses should have a clearly defined structure including an introduction accessible to scientists not expert in that particular field.

More information about this series at http://www.springer.com/series/8790

Henning Prüser

Scanning Tunneling Spectroscopy of Magnetic Bulk Impurities

From a Single Kondo Atom Towards a Coupled System

Doctoral Thesis accepted by
the University of Göttingen, Germany

Author
Dr. Henning Prüser
IV. Physikalisches Institut
University of Göttingen
Göttingen
Germany

Supervisors
Dr. Martin Wenderoth
Prof. Rainer G. Ulbrich
IV. Physikalisches Institut
University of Göttingen
Göttingen
Germany

ISSN 2190-5053 ISSN 2190-5061 (electronic)
ISBN 978-3-319-36051-5 ISBN 978-3-319-06385-0 (eBook)
DOI 10.1007/978-3-319-06385-0

Springer Cham Heidelberg New York Dordrecht London

Printed on acid-free paper

Springer is part of Springer Science+Business Media (www.springer.com)

Supervisors' Foreword

Eighty years ago, in 1934, the first experimental observation of the Kondo effect was reported. Since then Kondo effect and related phenomena have been thoroughly investigated in experimental and theoretical studies. One of the great hallmarks was the first descriptive interpretation, developed by Jun Kondo, giving this effect to its name. Magnetic impurities, i.e., elements with partially filled d- or f-orbitals, have localized spins that interact with the surrounding conduction electrons of a nonmagnetic metal. This interaction leads to a collective screening of the impurity spin below a characteristic temperature, the Kondo temperature. Below this temperature, such systems exhibit strong electronic correlations that give rise to various fascinating phenomena beyond the single particle description.

In 1998, the groups of Mike Crommie and Wolf-Dieter Schneider independently observed the Kondo effect at single magnetic impurities on a surface using scanning tunneling microscopy (STM) and scanning tunneling spectroscopy (STS). They showed that the central spectroscopy feature, the so-called many-body Abrikosov–Suhl resonance close to the Fermi-Energy, appears in current–voltage characteristics. Their pioneering works and new experiments performed at the same time at quantum-dot devices led to a revival of Kondo physics, as it was called by Leo Kouwenhoven and Leonid Glazman in 2001. Since then, the Kondo effect has been observed in a large variety of magnetic molecules and atoms on nonmagnetic surfaces.

This thesis describes a new experimental approach to investigate the Kondo effect of single magnetic impurities in the bulk as well as the interaction of selected dimers buried well below the surface of a Cu(100) crystal by means of low temperature STM and STS.

It is based on an observation made by Alexander Weismann: the possibility to investigate individual atoms below metal surfaces by STM. Impurities that act as local scattering centers modify the local density of states. For impurities located only few layers below the surface this modification is observable as an interference pattern at the surface. The STM as a surface sensitive tool gives access to true bulk properties.

Though experimental signals are relatively small and this approach is challenging, the investigation of bulk impurities offers two advantages. First, while STM and STS studies of magnetic molecules or atoms on noble metal surfaces show a spatially localized spectral feature, the corresponding results on embedded impurities exhibit complex spatially dependent signatures. Most important, this data give access to the mapping of the scattering phase caused by the Kondo resonance. As a result of this thesis and based on a detailed analysis of spatially resolved current–voltage-characteristics, an advanced description of the Kondo resonance as observed in STS was proposed. Second, bulk impurities are well described by theory, as the STM can be seen as a noninvasive probe. Indeed, theoretically proposed universal properties of a single-impurity Kondo system were found.

Based on a deeper understanding of the spectral signatures of a single impurity Kondo system, this thesis presents a first step toward the experimental study of coupled bulk Kondo systems, often called "dense" systems. While accepted models for single-impurity Kondo physics exist, a detailed knowledge of the transition from systems where the impurities can be treated individually toward dense systems is still lacking. Like for the single impurity Kondo model, the impurities are coupled to the surrounding conduction electrons, but also to each other by a magnetic exchange interaction. The latter can be due to an indirect coupling to the band electrons (RKKY interaction) or through other intermittent degrees of freedom (e.g., super- or double exchange). A qualitative picture can be derived by considering these two competing effects. While the single impurity Kondo effect tries to screen each impurity spin independently, the magnetic interaction between the impurities favors a nonlocal correlation between the localized spins. Antiferromagnetic as well as ferromagnetic coupling mediated by the RKKY interaction has been found in this work for different dimer configurations in the bulk.

In perspective, the interplay between the Kondo effect and the RKKY-interaction is a key element of the emergence of strongly correlated electron physics in many materials. A system of two magnetic impurities in a nonmagnetic metal, which is experimentally investigated in this thesis, is the smallest conceivable system including all ingredients for correlated electron physics in soids.

Dr. Henning Prüser completed his Ph.D. at the IV Physikalisches Insititut, Geog-August-University Göttingen in February 2013, under our supervision. In close collaboration with our colleagues from theory, two main results of his Ph.D. thesis have already been published in two scientific articles, a third one is in process.

(i) "Long-range Kondo signature of a single magnetic impurity", Nature Physics 7, 203 (2011)
(ii) "Mapping Itinerant Electrons around Kondo Impurities", Phys. Rev. Lett. 108, 166604 (2012)

In summary, this thesis can be thought of as a bottom approach toward better understanding of larger systems of strongly correlated electron systems.

Göttingen, May 2014

Dr. Martin Wenderoth
Prof. Rainer G. Ulbrich

Contents

Abbreviations and Symbols

Abbreviations

AC	Alternating current
AES	Auger electron spectroscopy
AFM	Antiferromagnetic
DC	Direct current
DFT	Density functional theory
DSP	Digital signal processor
FM	Ferromagnetic
HWHM	Half-width at half maximum
IPES	Inverse photoemission electron spectroscopy
KKR	Korringa–Kohn–Rostoker
LCAO	Linear combination of atomic orbitals
LDA	Local density approximation
LDOS	Local density of states
LEED	Low energy electron diffraction
LSP	Line-shape parameter ϕ
ML	Monolayer
NRG	Numerical renormalization group
PES	Photoemission electron spectroscopy
RKKY	Ruderman–Kittel–Kasuya–Yosida
SIAM	Single-impurity Anderson model
STM	Scanning tunneling microscope/microscopy
STS	Scanning tunneling spectroscopy
TIKM	Two-impurity Kondo model
UHV	Ultra high vacuum $p < 10^{-9}$mbar

Symbols

I	Tunneling current
I_{SP}	Set point current
V	Sample bias voltage
V_{SP}	Set point voltage
V_m	Modulation voltage amplitude added to the sample bias
ω_m	Modulation frequency
R_T	Tunneling resistance $R_T = V_{SP} I_{SP}^{-1}$
p	Vacuum pressure
n	Charge density
ρ^S	Local density of states of the sample
ρ^T	Local density of states of the tip
Φ^S	Work function of the sample
Φ^T	Work function of the tip
Φ	Apparent barrier height $\Phi = \left(\Phi^{\mathrm{T}} + \Phi^{\mathrm{S}}\right)/2$
κ	Decay constant
x	Lateral tip position, normally along the [010]-direction
y	Lateral tip position, normally along the [001]-direction
z	Tip-sample distance
z^*	Tip-sample distance of the unperturbed surface
d	Impurity position below the Cu(100) surface
r	Distance to an impurity $r = \sqrt{d^2 + x^2 + y^2}$
R	Interatomic distance of two iron atoms building a dimer
E_F	Fermi energy
ε	Energy with respect to the Fermi energy
$\varepsilon_{d,f}$	Energetic position of the d-, f–orbital
ε_k	Kondo resonance position
T_k	Kondo temperature $T_k \propto \Gamma \propto \Delta_k$
Δ_k	HWHM of the Kondo resonance $\Delta_k \propto T_k \propto \Gamma$
Δ_L	HWHM of the Lorentzian approximation $\Delta_L = \Gamma_L$
Δ_F	HWHM of the Frota approximation $\Delta_F = 2.54\,\Gamma_F$
ϕ	Line-shape parameter of the Kondo signature
Γ	Kondo resonance width according to the phenomenological fit formula $\Gamma = \Gamma_{\mathrm{F}} \propto k_B T_k \propto \Delta_k$
Γ^*	Kondo resonance width of the two-impurity Kondo model
Γ_L	Parameter defining the resonance width for the Lorentzian approximation $\Gamma_L = \Delta_L$
Γ_F	Parameter defining the resonance width for the phenomenological form found by Frota and coworkers $\Gamma_F = 0.39\,\Delta_L$
U	Coulomb repulsion of the localized orbital
J_{ex}	Exchange coupling between two localized spins
J	Exchange coupling between localized spin and conduction electrons
v_G	Group velocity

k	Wave vector $k = 2\pi\lambda^{-1}$
k_F	Fermi wave vector $k_F = 2\pi\lambda_F^{-1}$
k_{FW}	Friedel wave vector $k_{FW} = 2\pi\lambda_{FW}^{-1} = 2k_F$
λ	Wave length $\lambda = 2\pi k^{-1}$
λ_F	Fermi wave length $\lambda_F = 2\pi k_F^{-1}$
λ_{FW}	Friedel wave length $\lambda_{FW} = 2\pi k_{FW}^{-1} = \lambda_F/2$
dI_0/dV	Differential conductance of the free surface
dI/dV	Differential conductance of the tunneling current
$\Delta dI/dV$	Background subtracted differential conductance $= dI/dV - dI_0/dV$
m_e	Electron mass 9.11×10^{-31}kg
k_B	Boltzmann-constant 8.62×10^{-5}eV/K
$\hbar$	Reduced Planck constant 6.58×10^{-16}eV s

Chapter 1
Introduction

The Kondo effect, known for nearly 80 years is one of the first discovered correlation phenomena in solid state physics [1]. The phenomenon arises due to the presence of magnetic elements with partially filled d-, or f-orbitals in a non-magnetic host metal. As a result, a variety of low temperature thermodynamic anomalies can be observed, which are summarized by the term Kondo effect. The most prominent macroscopic hallmark is the resistance minimum at low temperature found for metals with magnetic impurities. From a microscopic point of view, the magnetic atoms have localized spins and thus internal degrees of freedom, which can interact with the surrounding conduction electrons of the metal. For low temperature the impurity spin is effectively screened by the conduction electrons. Although the phenomenon is well-known for a long time, the theoretical descriptions as well as the experimental investigations are challenging subjects of modern solid state research. The transition in particular from systems where the impurities can be treated individually towards dense systems is of fundamental interest. In the latter case where the density of magnetic impurities is sufficiently high, they can no longer be treated individually. The interplay between both the interaction between the impurities and the interaction with the conduction electrons is expected to provide the driving force for the emergence of strongly correlated electron physics in many materials. Especially materials where the local moments are arranged in a periodic array, so-called heavy-fermion systems [2][1] are widely studied during the last decades. In general, the correlations in the electronic system give rise to physics beyond the single particle picture. A correct description of localized moments in a metal requires a full many-body treatment.

Experimentally, strongly correlated electron systems have been investigated mainly by macroscopic transport (electrical resistivity) and thermodynamic measurements (specific heat, magnetic susceptibility and thermal expansion) [2, 3] or

[1] Since the localized magnetic moments and their correlation to the conduction electrons may give rise to electron quasi–particle masses up to thousand times higher than the bare electron mass, these materials are called heavy–fermion systems.

H. Prüser, *Scanning Tunneling Spectroscopy of Magnetic Bulk Impurities*,
Springer Theses, DOI: 10.1007/978-3-319-06385-0_1

high-resolution photoemission electron spectroscopy (PES) and inverse photo emission spectroscopy (IPES) [4]. As a result of their limited spatial resolution these techniques always probe a very large ensemble of impurities. Using these methods only a global characterization and no local information of the material can be obtained. This is in contrast to theoretical calculations considering only finite microscopic systems or systems of reduced dimensionality. This discrepancy often allows only comparing universal features and no system specific details. In order to investigate details and to further understand the impact of defects or disorder on the properties of strongly correlated systems, experimental tools with high spatial and spectral resolution are required.

The rapidly developing field of scanning tunneling microscopy (STM) and spectroscopy (STS) paved the way to tackle single-impurity [5] and two-impurity Kondo physics [6–10]. Since the first STM studies in 1998 [11, 12] showing Kondo fingerprints (see Fig. 1.1) many experiments on magnetic atoms and molecules *on metal surfaces* have been carried out.

With its unprecedented control the STM technique allows for the first time to investigate phenomena that occur on very small length and energy scales. It turns out that in an STM experiment the Kondo signature cannot be characterized by only one single feature, e.g. a peak or a dip, but also by asymmetric line shapes. This is in contrast to PES and IPES showing always a peak-like structure in the spectrum.

Studies which investigate the dependence of the Kondo signature on the lateral distance showed that the Kondo signature rapidly vanishes with increasing distance from the atom [12–14]. This is in contrast to theory [15–17], which predicts a spatially extended Kondo feature. In addition, the interaction of two neighboring Kondo impurities only plays an important role for an interatomic distance of a few angstroms [6–9]. For a greater distance a coupling between the impurities can be achieved when the magnetic atoms are connected by a non-magnetic atomic chain [10]. The missing spatial extension of the Kondo signature and the weak interaction between the magnetic atoms and molecules on noble metal surfaces make it rather complicated to compare the results to macroscopic measurements of strongly correlated systems.

A new experimental approach to study single impurities was introduced by Alexander Weismann and Colleagues [18]. It turns out that it is generally possible to investigate individual atoms *below metal surfaces* by STM [19, 20]. The impurities act as local scattering centers modifying the local density of states (LDOS) around the impurity. The interference pattern at the surface can be measured and analyzed as function of applied bias voltage.

Following the idea of embedded impurities, in this thesis silver (Ag), iron (Fe) and cobalt (Co) atoms buried below a Cu(100) surface are investigated using low temperature STM and STS ($T = 6\,\mathrm{K}$). In Fig. 1.2 a sketch of the experiment is depicted. Single Fe and Co atoms belong to the 3d transition metals and both exhibit a partially filled d-orbital. From macroscopic measurements [3], showing the Kondo effect, it is well-known that the local moment of Fe and Co survives in

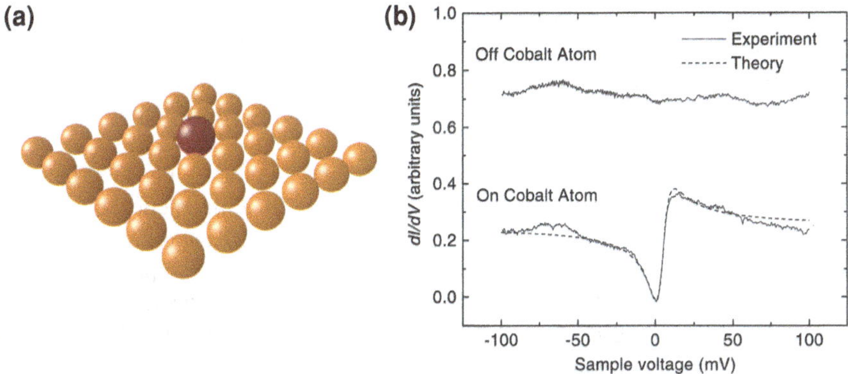

Fig. 1.1 Manifestation of the Kondo effect on single magnetic adatoms. **a** Sketch of a single magnetic impurity located on a crystalline surface. **b** Experimental STS results for single cobalt atoms on an Au(111) surface. In the spectroscopic data, a dip-like feature appears above the cobalt atom, which is interpreted as a fingerprint of the Kondo effect. In contrast, the bare Au(111) surface shows no signature in the differential conductance around zero sample bias voltage. From [12]. Reprinted with permission from AAAS

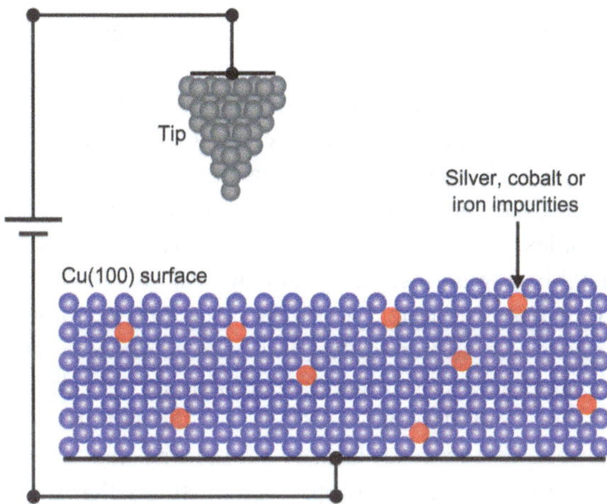

Fig. 1.2 A sketch of the experimental part presented in this thesis. Silver, cobalt and iron impurities embedded below a Cu(100) surface are investigated by means of low temperature scanning tunneling microscopy and spectroscopy

copper. Silver atoms are non-magnetic and represent a control experiment to verify that the observed features of Fe and Co can be related to their localized moment.

This thesis focuses on single impurity-Kondo physics and *the transition towards coupled Kondo systems* in real space. Although the STM is a surface sensitive method the results can be directly related to bulk properties, which is in contrast to adatom systems. Therefore, the experimental findings are not only relevant for artificial nanostructures but also may be relevant for bulk systems. In particular, the results obtained on the nano scale can be compared for the first time to the macroscopic properties of the material.

The work is structured as follows: In Chap. 2 the sample preparation and the low temperature STM and its underlying theory are introduced. The signatures caused by the *bulk impurities* are strongly related to the band structure of the copper crystal. The local density of states in the vicinity of an impurity will be discussed in Chap. 3.

Measurements of single isolated *magnetic* Co and Fe atoms embedded below a Cu(100) surface are presented in Chap. 4 showing that magnetic sub-surface impurities offer a hitherto unobserved spatially extended Kondo signature. The main idea for the description of Kondo bulk impurities is the separation of the electron propagation, determined by the host properties, and the scattering characteristics of the impurity, which is strongly affected by the Kondo effect. A simulation combining band structure calculations and many-body calculations[2] is able to reproduce the main experimental findings. A detailed analysis of spectroscopic data as function of lateral tip position and impurity distance below the surface reveals an oscillation of the line shape of the Kondo signature, which has been predicted theoretically in the past but so far was not observed in the experiment.

A control experiment to verify that the observed features for magnetic Fe and Co are really due to their localized magnetic moment is presented in Chap. 5. Here *non-magnetic* Ag impurities are considered. While for magnetic Co and Fe impurities the signature strongly depends on the applied bias voltage for Ag atoms only a minor dependence is found—the scattering behavior seems to be nearly energy independent. Nevertheless, spectroscopic data reveal an unexpected small kink around zero bias voltage. Up to now the physical origin of this feature not related to Kondo physics is not fully clear.

In Chap. 6 systems with higher Fe doping concentrations are considered. These samples allow investigating not only single isolated impurities but also two neighboring atoms. Fe *dimers* show a considerably different spectral signature as compared to single impurities taken on the same sample. The observed Kondo features are strongly altered or even suppressed due to the presence of the second Fe atom. It turns out that the behavior strongly depends not only on the interatomic distance but depends crucially on the atomic configuration of the dimer with respect to the host lattice. Comparing the experimental findings with recent first-principles calculations[3] demonstrates that the properties of two neighboring iron atoms can be related to the indirect Ruderman-Kittel-Kasuya-Yosida (RKKY) interaction. The calculation is able to reproduce both the strength as well as the strong directionality of the magnetic interaction found in the experiment.

[2] The simulation and the many-body calculations originate from collaboration with Piet E. Dargel, Robert Peters and Thomas Pruschke from the Institute for Theoretical Physics at the University of Göttingen, Germany.

[3] The first-principles calculations originate from collaboration with Mohammed Bouhassoune and Samir Lounis from the Institute for Advanced Simulation at the Research Centre Jülich, Germany.

In Chap. 7 the experimental findings are compared to previous works on adatom systems followed by a discussion of the perspective for magnetic bulk impurity systems. Finally, in Chap. 8 the results of this thesis are summarized.

References

1. Hewson AC (1993) The Kondo problem to heavy fermions. Cambridge University Press, Cambridge
2. Stewart GR (1984) Rev Mod Phys 56:755
3. Daybell MD, Steyert WA (1968) Rev Mod Phys 40:380
4. Reinert F, Hüfner S (2005) New J Phys 7:97
5. Ternes M, Heinrich AJ, Schneider W-D (2009) J Phys: Condens Matter 21:053001
6. Chen W, Jamneala T, Madhavan V, Crommie MF (1999) Phys Rev B 60:R8529
7. Jamneala T, Madhavan V, Crommie MF (2001) Phys Rev Lett 87:256804
8. Madhavan V, Jamneala T, Nagaoka K, Chen W, Li J-L, Louie SG, Crommie MF (2002) Phys Rev B 66:212411
9. Wahl P, Simon P, Diekhöner L, Stepanyuk VS, Bruno P, Schneider MA, Kern K (2007) Phys Rev Lett 98:056601
10. Néel N, Berndt R, Kröger J, Wehling TO, Lichtenstein AI, Katsnelson MI (2011) Phys Rev Lett 107:106804
11. Li J, Schneider W-D, Berndt R, Delley B (1998) Phys Rev Lett 80:2893
12. Madhavan V, Chen W, Jamneala T, Crommie MF, Wingreen NS (1998) Science 280:567
13. Knorr N, Schneider MA, Diekhöner L, Wahl P, Kern K (2002) Phys Rev Lett 88:096804
14. Henzl J, Morgenstern K (2007) Phys Rev Lett 98:266601
15. Újsághy O, Kroha J, Szunyogh L, Zawadowski A (2000) Phys Rev Lett 85:2557
16. Schiller A, Hershfield S (2000) Phys Rev B 61:9036
17. Plihal M, Gadzuk JW (2001) Phys Rev B 63:085404
18. Weismann A (2008) Scanning tunneling Spectroscopy on subsurface magnetic atoms in copper: electron focusing and Kondo effect. PhD thesis, Georg-August-Universität zu Göttingen
19. Weismann A, Wenderoth M, Lounis S, Zahn P, Quaas N, Ulbrich RG, Dederichs PH, Blügel S (2009) Science 323:1190
20. Lounis S, Zahn P, Weismann A, Wenderoth M, Ulbrich RG, Mertig I, Dederichs PH, Blügel S (2011) Phys Rev B 83:035427

Chapter 2
Experimental Setup and Background

The experimental task of this thesis is to prepare magnetic and non-magnetic bulk impurities and to investigate these systems by scanning tunneling microscopy (STM). In the first part of this chapter the sample preparation will be described. The main experimental instrumentation, the low temperature STM as well as the standard theory for the interpretation are introduced in the second part.

2.1 Sample Preparation

The basic idea of preparing single sub-surface atoms is very simple. Starting with a clean Cu(100) single crystal a thin alloy containing the host and the impurity material is epitaxially grown onto the substrate by using two electron beam evaporators. In order to get a well-defined sample system, the preparation has to meet several requirements. For the investigation of single atoms a low concentration of impurities is sufficient. Therefore, the host evaporator operates at moderate rates (1–2 ML/min, monolayer per minute), whereas the second one is kept at very low deposition rates (0.1 ML/min). To avoid unwanted contamination the whole preparation is performed under ultra-high vacuum (UHV) conditions ($p < 10^{-9}$mbar) using high purity materials and a carefully cleaned substrate. Furthermore, the resulting surface should provide large atomically flat areas to get a well-defined tunneling geometry.

2.1.1 UHV Preparation Chamber

The sample preparation is carried out in a mobile UHV-chamber. In combination with a valve and a turbo-molecular pump this chamber allows an immediate transfer of the sample to various existing UHV-chambers for example the

H. Prüser, *Scanning Tunneling Spectroscopy of Magnetic Bulk Impurities*,
Springer Theses, DOI: 10.1007/978-3-319-06385-0_2

STM-chamber or the characterization chamber without breaking the vacuum. After 48 h bake-out at 150 °C a base pressure of $p < 5 \times 10^{-11}$mbar is achieved.

For the sample preparation the chamber is equipped with three independent operating electron beam evaporators, an ion sputtering system and a pyrometer for non-contact temperature measurement. The sample is mounted below a liquid nitrogen (LN_2) tank which allows cooling of the sample. Using a tungsten filament the sample can be heated from the back side of the crystal. With this setup substrate temperatures from 100 K up to 1,300 K can be achieved. Additionally, the sample holder possesses six electrical contacts (multi-segment sample holder). The electrical contact is necessary to apply a high voltage to the sample in order to perform electron beam heating and to suppress ion bombardment during evaporation. Furthermore, the electrical contact is used to ground the sample during sputtering. The current directly gives rise to the amount of argon ions hitting the sample per second. A detailed description of the sample preparation chamber can be found in [1, 2].

2.1.2 Single Crystal Preparation and Characterization

As substrate material a single crystal of copper is used in this work, which is commercially available from MaTeck, Jülich. The crystal is 1 mm thick and has a hat-like shape with diameters of 8.5 mm and 6.0 mm, respectively. The surface is oriented normal to the [100] direction with an accuracy better than 0.1 % and is polished to a roughness smaller than 30 nm. According to the specification the purity is better than 99.9999 %.

In order to get a well-defined epitaxial growth of the copper alloy one needs a clean and atomically flat substrate. In case of the Cu(100) single crystal substrate this goal can be achieved by cycles of argon ion bombardment at room temperature followed by electron beam heating. Typically ten cycles are enough to achieve a clean surface. The ion bombardment leads to sputtering of the surface as well as to argon implantation in the surface region. Both effects lead to a distorted crystalline structure at the surface. The crystalline structure is recovered by a short annealing step. The temperature must be high enough or the duration long enough that all implanted argon atoms diffuse to the surface and leave the sample. Moreover, during the annealing bulk defects segregate towards the surface. They are removed in subsequent sputtering steps. In the following a detailed description of the methods for the Cu(100) single crystal preparation will be given.

2.1.3 Sample Holder

The sample holder has to fulfill several requirements: on the one hand a good thermal contact to the cryostat is essential. On the other hand to heat only the

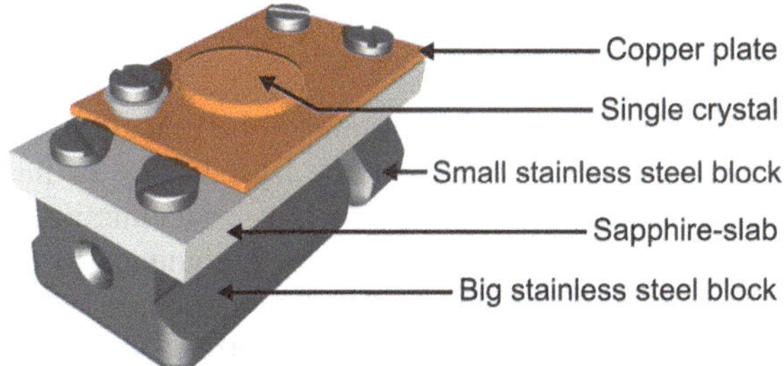

Fig. 2.1 Design of the low temperature sample holder. The Cu(100) crystal is fixed onto the sapphire slab by a copper plate. An electrical contact to the crystal is provided by the backward small stainless steel block

crystal and not the whole sample holder an electrical insulation is necessary for the application of electron beam heating. To meet these requirements the sample holder consists of two different sized dovetail-shaped blocks made of stainless steel. The blocks are hold together by a sapphire-slab on which the copper crystal is placed. Sapphire is an electrical insulating material which has a high thermal conductivity. To beware of contamination by secondary sputtering effects the single crystal of copper is pressed with caution onto the sapphire-slab by a copper plate or copper-covered molybdenum plate. The complete design of the sample holder is sketched in Fig. 2.1.

The crystal can be electrically contacted by the small stainless steel block. A hole in both, the big block and the sapphire-slab gives access to the back side of the crystal. For the electron beam heating process a tungsten wire can be inserted in this opening. The geometry of the sample holder and the hat-like shape of the crystal avoid direct trajectories between sample holder and the crystal surface. Therefore, a contamination by secondary processes during argon sputtering is suppressed.

2.1.4 Argon Sputtering

For removing surface material a sputtering source (IQE11/35 by SPECS, Berlin) is used. The general concept of sputtering can be summarized as follows: Argon atoms, entering the preparation chamber through a variable leak valve, are ionized by an electron beam and accelerated to the sample by a high voltage. By tuning the pressure in the chamber and by measuring the ion flux on the sample the intensity of the argon ion beam is adjustable. Typical values for the current densities are $0.81–2 \times 10^{-2}\,\mathrm{A\,m^{-2}}$ with 700 eV ion beam energy. At the beginning of the preparation the sputtering goes on for one hour to remove the oxide film and structures coming from prior experiments. After the sample is sputtered for one hour, the sputtering time is reduced to 20 min for all following cleaning cycles.

2.1.5 Annealing

The sample heating is realized by electron bombardment. A tungsten wire is brought into close contact to the backside of the crystal using a z-micrometer drive. Filament currents in the order of 5–7 A lead to a thermal emission of electrons that are accelerated to the crystal by applying a positive voltage to the sample. This setup allows powers up to 50 W resulting in temperatures up to 1,300 K. The temperature can be monitored by a pyrometer (IM120 by IMPAC, Frankfurt) in the range from 430 K to 1,570 K. For the emissivity of the clean Cu(100) surface a value of 11 % is proposed [1]. During the annealing process the crystal is heated up by a rate of approximately 50 K/min to a final temperature of 700 K. When the temperature is reached, it is kept constant for 10 min followed by a step by step cool down extended over 10 min. For the last cycle, before starting the thin film epitaxy the annealing time is reduced to 5 min to avoid segregation of bulk defects.

2.1.6 Substrate Characterization

The single crystal preparation procedure is tested by the non-local surface sensitive methods low energy electron diffraction (LEED) and Auger electron spectroscopy (AES). These two methods allow investigating the crystalline quality of the sample and the chemical composition of the surface after the preparation, respectively.

In Fig. 2.2a the AES spectrum before the cleaning procedure is shown. By comparing the signature with reference spectra [3] one observes that besides copper (Cu) also oxygen (O), carbon (C) and molybdenum are present at the beginning of the sample preparation. After three cycles of sputtering and heating the oxygen and the carbon signature decreases while the intensity of the copper peaks increases. Finally, after nine cycles no signature of carbon and oxygen is observed. This shows that the cleaning procedure works very well resulting in a defect free surface. Only peaks referred to molybdenum and copper are found in the spectrum. The peaks belonging to molybdenum are due to defocussing of the electron beam which hits not only the crystal but also the sample holder.

The crystalline quality is analyzed by the LEED diffraction pattern, see Fig. 2.2b. It shows sharp peaks resembling the diffraction characteristics of an fcc-crystal structure. The high quality of the cleaned Cu(100) substrate is confirmed by a high resolution STM topography (see Fig. 2.2c) revealing large and homogeneous terraces, which are separated by atomic steps.

2.1.7 Epitaxy of Diluted Cu Films

To get well-defined alloys only high purity materials are used in this work. The materials were purchased from Alfa Aesar and have a purity of 99.995 % (Co, Fe),

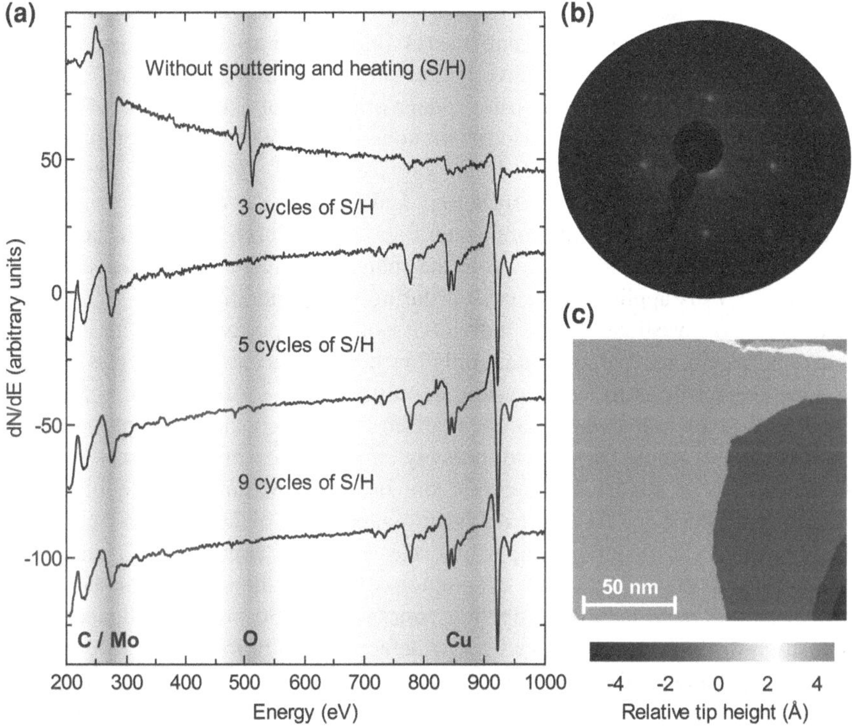

Fig. 2.2 Characterization of the prepared Cu(100) surface. **a** AES spectra (*shifted vertically for clarity*) before and after several cycles (3, 5, 9) of sputtering and heating. Characteristic energy intervals for the atomic species C, O and Cu are highlighted by color. **b** After the cleaning procedure the LEED diffraction pattern taken at 170 eV reveals sharp peaks. **c** In the STM topography (150 nm × 150 nm, 1 V, 50 pA) a flat surface is observed which offers large and defect free terraces

99.9999 % (Ag) and 99.9999 % (Cu). Before the deposition, the evaporators are outgassed for roughly 2 h. This means that they are kept at high flux rates (compared to the actual thin film epitaxy) until the pressure in the preparation chamber is constant.

After manufacturing a clean and flat Cu(100) surface, thin films of diluted copper are deposited onto the clean substrate. Therefore, electron beam evaporators (e-beam by TECTRA, Frankfurt) are used. Electrons which are emitted by a tungsten filament are accelerated due to high voltage to the material to be evaporated. The evaporant, which is either a rod (Co and Fe) or located in a tungsten crucible (Cu and Ag), is heated up and goes into the gas phase. A small percentage of these vapor-phase atoms are ionized by the electron beam. A fraction of the ionized atoms can be captured and used to control the material flux. A detailed description of the evaporators, their functionality as well as their close-loop control is given in [1, 2, 4–6]. In the following only the most important facts and parameters for the preparation of Ag-, Co- and Fe-diluted copper films will be discussed.

It turned out that for the deposition of copper one has to take care of positively charged atoms. Although they are useful for the control of the flux they are repelled by the crucible and directly accelerated to the sample. The ion flux, measured directly at the sample using a deposition rate of 1 ML/min is 0.4×10^{-2} A m^{-2}. This value and the energy of the ions (800–1,400 eV) are comparable to the argon bombardment used for cleaning the crystal. As a result, the ionized atoms have an impact on the morphology and the growth process for thin copper films. A detailed study of the influence of sputtering during growth of copper on Cu(100) can be found in [6]. To avoid the sputtering effect a high positive voltage (800–1,400 V) is applied to the crystal during the evaporation.

In order to investigate single individual isolated impurity atoms one has to be able to prepare alloys which contain only low defect concentrations of less than 0.1 %. One possibility is to use very low deposition rates for the impurity material. The lowest stable rates which can be realized for iron, cobalt and silver are of the order of 0.1 ML/min. To achieve impurity concentration of less than 0.1 %, this implies a copper deposition rate of more than 100 ML/min. These high rates cannot be reached for the electron beam evaporation of copper. To obtain low concentrations for moderate copper flux rates computer controlled stepper motors are attached to the shutter of the evaporator containing the impurity material. The stepper motors provides shutter opening times down to 50 ms. Using this setup the deposition of thin copper films with low defect concentrations down to 0.01 % is possible by using moderate copper fluencies of about 0.5–1 ML/min. In the following the preparation of Co-, Fe- and Ag-diluted copper films is described in detail.

2.1.8 Co in Cu(100)

After the cleaning procedure six monolayers of copper were deposited onto the substrate using a deposition rate of 1 ML/min. In a next step, under continuous copper deposition the cobalt shutter is opened for 100 ms every minute for 12 min. At the same time, the deposition rate for Co is stabilized at 0.13 ML/min. This value theoretically corresponds to a cobalt concentration of $c_{Co} = 0.02$ % and is in good agreement to the concentration that is observed in the experiment [6]. Finally, two layers of pure copper are deposited. The whole epitaxy is carried out at a substrate temperature[1] $T_S = 420\,\mathrm{K}$ to ensure a layer by layer growth resulting in a very flat surface [6].

[1] According to calibration measurements, a constant heating power of 3 W corresponds to a sample temperature of $T_S = 420\,\mathrm{K}$.

2.1.9 Fe in Cu(100)

First attempts trying the same recipe for the preparation of iron doped copper films result in a high number of iron clusters located in the surface layer. Nearly all Fe atoms segregate to the surface layer for the same growth conditions as for Co ($T_S = 420\,\mathrm{K}$). A way to reduce the mobility of the iron atoms during epitaxy is to go to lower temperatures. Therefore, the substrate is cooled down to 100 K prior to deposition. After doing this, nine monolayers of copper are evaporated using a deposition rate of 0.5 ML/min. Then for 18 min every 2 min the iron shutter is opened for 160 ms. The deposition rate is well stabilized at 0.2 ML/min. Finally, two monolayers of copper were deposited on top. The low temperature growth hinders the iron atoms to segregate to the surface but has the drawback of producing a very rough surface not suitable for an STM experiment. To overcome this shortcoming, the sample is flushed to a temperature of 470 K for 5 s. After this the electron beam heating is rapidly ceased. The short annealing step is sufficient for surface diffusion of copper atoms and results in a flat surface with $50\,\mathrm{nm} \times 50\,\mathrm{nm}$ atomic step free terraces. Although the concentration of the iron atoms embedded in the copper crystal should be in principle the same as in the cobalt case described above, only a fraction of the concentration of the impurities is located below the surface. There are still iron clusters in the first atomic layer of the surface indicating that there is still room for sample preparation improvements.

2.1.10 Ag in Cu(100)

The silver-doped copper film is prepared in the same manner using a low temperature deposition. After eight monolayers of copper are deposited at 1 ML/min, ten monolayers of a 0.02 % silver-copper alloy are grown. Then, two layers of copper are grown on top. Finally, the sample is annealed to 470 K followed by a fast cool down procedure.

2.2 Scanning Tunneling Microscopy

The scanning tunneling microscope (STM) was invented by G. Binnig and H. Rohrer in 1982 at the IBM research center in Zürich [7, 8]. It is based on the concept of tunneling. This quantum effect causes a current flow between a metal tip and a conductive sample if they are brought into close proximity to each other with a bias voltage applied simultaneously. For a given tip-sample system the resulting tunneling current is a function of the applied bias voltage, the tip-sample distance and the lateral tip position. Keeping the bias voltage and the lateral tip position fixed, the tunnel current depends exponentially on the tip-sample distance.

This fact can be utilized to stabilize the tip height by using a close-loop control and piezoelectric elements. Local information about the sample is gained by scanning the tip across the surface (x, y).

Since the tip is separated only a few angstroms from the surface, the STM technique requires excellent damping of internal (liquid nitrogen and helium) and external (building and noise) vibrational sources. Moreover, clean surfaces, stable and sharp tips and good electronic equipment are necessary for high quality STM measurements. An introduction to scanning tunneling microscopy and related techniques can be found in [9–11].

The experiments in this thesis were performed using a home built low temperature STM operating at pressures below $p < 5 \times 10^{-11}$mbar. The scanning unit is a Besocke-type piezo-unit [12] mounted on the bottom of a liquid helium (LHe) bath cryostat. The scanner and the sample support are thermally coupled to the helium bath via sapphire plates and indium foils. Small silver wires interconnect the tip and the cryostat. A liquid nitrogen (LN_2) vessel surrounds the LHe-tank and STM-head shielding the heat radiation. An optical access is provided by windows in the nitrogen cup. These openings are covered by special glasses that absorb infrared radiation. This setup provides a sample and tip cooling and the possibility replacing both of them in-situ. The sample temperature is estimated around 6 K. Recent scanning tunneling potentiometry studies investigating the thermovoltage between tip and sample indicate that the temperature of the tip is slightly higher than 6 K [13]. Measurements considering the Cu(100) surface reveal an thermovoltage signal—the experimentally uncertainty of the applies bias voltage—with amplitude up to 0.3 mV. This value is small compared to the voltage ranges considered in this work and hence the thermovoltage is neglected in the following.

2.2.1 Standard STM Theory

There exists no complete theoretical formalism describing the tunneling process in all its diversity. The most widely used models start with a tunneling formalism proposed by Bardeen calculating the current from the sample and tip states of the unperturbed systems using first order perturbation theory [14]. Transition of electrons can only take place from filled sample states φ_ν^S into empty tip states φ_μ^T or from filled tip states into empty sample states. Applying a voltage V to the sample, the tunneling current $I(V)$ can be expressed as

$$I(V) = \frac{4\pi e}{\hbar} \sum_{\nu,\mu} \left[f\left(E_\mu^T - E_F^T\right) - f\left(E_\nu^S - E_F^S\right) \right] \left|M_{\nu,\mu}\right|^2 \delta\left(E_\nu^S - E_\mu^T - eV\right) \quad (2.1)$$

where $f(\varepsilon) = (1 + \exp(\varepsilon/k_B T))^{-1}$ is the Fermi-Dirac distribution and $M_{\nu,\mu}$ is the tunneling matrix element describing the transition probability from state φ_ν^S to state φ_μ^T. The Dirac delta function restricts the tunneling process to states fulfilling energy conservation. Under the assumption that the tunneling matrix element $M_{\nu,\mu}$ does only depend on the energetic position of the considered states, the summation of the states in Eq. (2.1) can be replaced by an integration over the density of states (DOS) of tip $\rho^T$and sample ρ^S.

$$I(V) = \frac{4\pi e}{\hbar} \int d\varepsilon \left[f\left(E_F^S - eV + \varepsilon\right) - f\left(E_F^T + \varepsilon\right)\right] \\ \times \rho^S\left(E_F^S - eV + \varepsilon\right) \rho^T\left(E_F^T + \varepsilon\right) \left|M\left(E_F^S - eV + \varepsilon, E_F^T + \varepsilon\right)\right|^2 \tag{2.2}$$

This formula links the tunneling current to the electronic properties of tip and sample described by the density of states. The distance dependence as well as the lateral dependence of the current are included in the matrix element M. For low temperature the Fermi-Dirac distribution can be approximated by a step function.

$$I(V) = \frac{4\pi e}{\hbar} \int_0^{eV} d\varepsilon \rho^S\left(E_F^S - eV + \varepsilon\right) \rho^T\left(E_F^T + \varepsilon\right) |M|^2 \tag{2.3}$$

Assuming further small voltages V the integral can be simplified. The tunnel current then reads

$$I(V) = \frac{4\pi e^2}{\hbar} V \rho^S\left(E_F^S\right) \rho^T\left(E_F^T\right) \left|M\left(E_F^S, E_F^T\right)\right|^2 \tag{2.4}$$

The model so far for the tunneling current is a consequence of Bardeen's tunneling formalism connecting the current to a transfer matrix element. In a next step, it is necessary to calculate this element for the special geometry of a sharp tip in front of a sample. In general, the shape and the electronic structure of the tip are unknown in the experiment. Tersoff and Hamann approximated the tip as an s-like wave function centered at the tip apex $\vec{r}^T$. They relate the tunnel current to the local density of states of the sample at the position of the tip [15, 16].

$$I(V) \propto V \rho^T(\varepsilon) \rho^S\left(\varepsilon, \vec{r}^T\right) \tag{2.5}$$

According to Tersoff and Hamann the tunneling current is proportional to the applied voltage, the tip density of states and the sample density of states at the tip position.

For STM measurements at higher voltages the above assumption of a constant energy-independent transfer matrix element is no longer valid. The matrix element changes for states of different energies. A complete energy-dependent description of this problem is rather complicated. A simplified model by Hamer's replaces the tunneling matrix element $|M|^2 \rightarrow T(\varepsilon, eV, z)$ by a transition probability $T(\varepsilon, eV, z)$ [17]. The current is then obtained by an integration of all possible energies where tunneling can occur.

$$I(V,x,y) = \frac{4\pi e}{\hbar}\int_0^{eV} d\varepsilon \rho^S(E_F^S - eV + \varepsilon, x, y)\rho^T(E_F^T + \varepsilon)T(\varepsilon, eV, z) \tag{2.6}$$

The transmission probability is calculated using the Wentzel-Kramers-Brillouin approximation and a trapezoid shape of the tunneling barrier.

$$T(\varepsilon, eV, z) \propto \exp\left(\frac{-2z\sqrt{2m_e}}{\hbar}\sqrt{\frac{\Phi^{\mathrm{T}} + \Phi^{\mathrm{S}}}{2} + \frac{eV}{2} - \varepsilon}\right) \tag{2.7}$$

The vacuum barrier is determined by the tip and sample work function Φ^T and Φ^S respectively, as well as by the tip-sample distance z. The model includes also the change of the barrier height due to the applied bias voltage V and the electron energy ε. Compared to the theory of Tersoff and Hamann, in the model of Hamer's the sample local density of states is considered at the lateral surface position $(x, y, 0)$ and not at the tip position $\vec{r}^T = (x, y, z)$.

2.2.2 Constant Current Topography

In the constant current mode, a voltage V is applied between tip and sample. The tip height is adjusted so that a constant set point current I_{SP} flows between tip and sample. The tip is scanned laterally (x, y) over the sample line by line using piezo-elements. In the meantime a closed-loop feedback control circuit adjusts the tip height z in such a way that at each point the set point current is reached. In each line the tip displacement is recorded at discrete points, providing a contour $z(x, y)$ of constant current. For small voltages V the area $z(x, y)$ can be interpreted according to Tersoff and Hamann (Eq. 3.9) as a surface of constant local density of states at the Fermi energy. In case of higher voltages, the area $z(x, y)$ has to be interpreted with respect to the model of Hamer as a surface of constant integrated density of states in the interval $[E_F, E_F + eV]$. In both cases, the height profile is a superposition of structural and electronic properties of the sample. For example in a large-scale topography (see Fig. 2.2c) sharp jumps with a height of $\Delta z = 180\,\mathrm{pm}$, monoatomic steps of the Cu(100) surface, are found which are related to the morphology of the sample.

In this thesis the electronic structure caused by sub-face atoms is of interest. This signature is in general much smaller than a monoatomic step or an adatom on a clean surface. For instance the raw data (only an average plane is subtracted) of a topography measurement (4 nm × 4 nm, 128 pixel × 128 pixel) of two sub-surface iron atoms is shown in Fig. 2.3a. One can see an interference pattern with amplitude of only a few picometers. The local density of states is changed due to the presence of the sub-surface atoms resulting in a standing-wave pattern at the surface. Above areas of reduced electron density in the energy interval

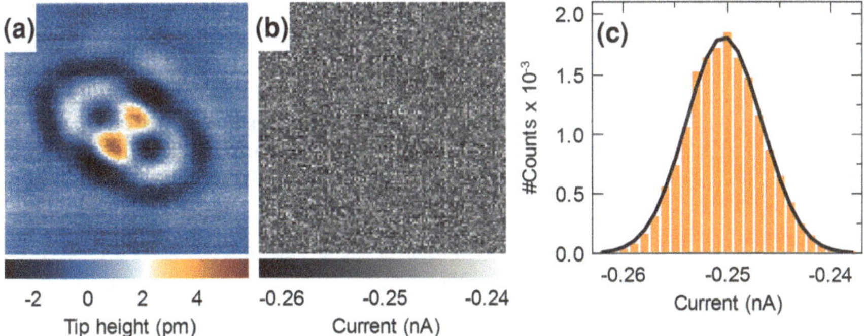

Fig. 2.3 Characterization of the constant current mode. **a** Topography (4 nm × 4 nm, −31 mV, 0.25 nA, raw data) of two sub-surface Fe atoms. The corresponding current measurement is depicted in **b**, as current map and in **c**, as histogram. A Gaussian fit is marked in *black*

$[E_F, E_F + eV]$ the tip has to move closer to the surface. In contrast, the tip is retracted above regions of increased electron density. In order to check the quality and to avoid artifacts of the feedback loop not only the tip displacement but also the current is acquired in the topography mode. If the feedback loop is adjusted, the current map (Fig. 2.3b) shows only a small lateral variation with respect to the set point current I_{SP}. A histogram of the current map Fig. 2.3c reveals a standard deviation of $\sigma = 0.005\,\text{nA}$ which is 2 % of the current set point $I_{SP} = 0.25$ nA.

2.2.3 Multi-bias Topographies

The fastest way to investigate the electronic structure of the sample is to measure topographies of the same area for different applied voltages. If the work function of the sample is laterally constant, changes in the topography can be related to the integral LDOS of the sample. According to Eq. (2.6) the tunneling current is proportional to the integrated local electron density of states $\rho^S(\varepsilon)$ at the surface within $[E_F, E_F + eV]$. This quantity is weighted in the integral by the tip local density of states $\rho^T(\varepsilon)$ and the transmission probability $T(\varepsilon, eV, z)$. As a consequence, states with different energy contribute differently to the tunneling current. For energy intervals where the variation of the electronic properties of the tip and the transmission probability are negligible the measured contour is proportional to the average LDOS.

As the recording of a topography takes some time, it might be possible that the same lateral tip positions are not comparable for different topographies at different bias voltages due to drift effects or tip modifications. To overcome this problem one can use the multi-bias mode. In this mode every scan line is scanned for several tunneling parameter settings (V_n, I_n) before going to the next scan line. This mode provides contours $z_n(x, y)$ which are comparable at the same lateral position (x, y).

2.2.4 *dI/dV* Spectroscopy

One of the most powerful features of scanning tunneling microscopy is its capability to obtain spectroscopic data with high spatial and energetic resolution. Scanning tunneling spectroscopy (STS) provides information about the LDOS of the sample as function of the lateral position (x, y) and a certain energy ε. The LDOS can be approximately derived from the differential conductance $dI/dV(V)$ at a certain bias voltage V. Since the main results presented in this thesis are obtained by spectroscopic measurements, the description of this technique and its interpretation will be discussed in detail in the following section.

Taking the tunneling current expressed by Eq. (2.6) and using Leibnitz's rule, the derivative of the current gives three terms

$$\begin{aligned}\frac{dI}{dV}(V,x,y,z) \propto\ & \rho^S(x,y,eV)\rho^T(0)T(eV,V,z) \\ & + \int_0^{eV} d\varepsilon\rho^S(x,y,\varepsilon)\rho^T(\varepsilon-eV)\frac{dT(\varepsilon,V,z)}{dV} \\ & + \int_0^{eV} d\varepsilon\rho^S(x,y,\varepsilon)T(\varepsilon,V,z)\frac{d\rho^T(\varepsilon-eV)}{dV}\end{aligned} \tag{2.8}$$

While this equation is quite complicated, it can be simplified assuming a constant LDOS of the tip ($\rho^T = \text{const}$) and a constant transition probability ($T(\varepsilon, V, z) = \text{const}$). These assumptions are valid for small energies, i.e. ε is small compared to the work functions of tip and sample and $\rho^T(\varepsilon)$ is featureless compared to $\rho^S(\varepsilon)$. With these assumptions the second and the third term vanish and the derivative becomes

$$\frac{dI}{dV}(V,x,y) \propto \rho^S(x,y,eV) \tag{2.9}$$

For the investigation of sub-surface atoms the change of the LDOS $\Delta\rho^S(x,y,\varepsilon)$ due to the presence of the impurity in comparison to the unperturbed surface is of interest. This quantity can be obtained by measuring a reference spectrum of the free unperturbed surface $dI_0/dV(V)$ far away from the impurity. The background subtracted differential conductance $\Delta dI/dV(V)$

$$\frac{\Delta dI}{dV} = \frac{dI}{dV}(x,y,V) - \frac{dI_0}{dV}(V) \propto \Delta\rho^S(x,y,eV) \tag{2.10}$$

is proportional to the change in the LDOS at the position (x, y) in comparison to the unperturbed surface [18].

Technically, the tip is placed over an area of interest with a defined distance to the sample given by the stabilizing tunneling parameters (V_{SP}, I_{SP}). Then the feedback loop is temporarily switched off and a spectrum is recorded within a certain voltage range. After that, the feedback loop is reactivated and the tip moves to the next scan point. Experimentally, there are several possibilities to measure the differential conductance. One approach is to record an *I*(*V*)-curve at every scanning point. By using further data processing including averaging and numerical differentiation this provides a complete map of the differential conductance $dI/dV(x,\ y,\ V)$ as a function of lateral tip position $(x,\ y)$. Although the numerical approach is simple, the obtained spectroscopic data is usually very noisy. Because of this one has to average over a long time to obtain reasonable data. To get high resolution spectra on shorter time scales a lock-in technique has been developed and implemented into the existing system. The basic idea rest upon the observation that the noise of the tunneling current is not determined by white noise but rather by 1/f noise. The advantage of the lock-in technique is to measure the first derivative directly and to shift the signal from zero frequency (as it is the case for the numerical derivative) with a high noise level towards higher frequencies with a lower noise level.

2.2.5 Lock-in Technique

A lock-in method was implemented to obtain spectroscopic data with high signal to noise ratio. In order to control all necessary parameters of this technique and to integrate the lock-in amplifier into the existing electronic setup a digital signal processing (DSP) board (ADwin-light-16 by Jäger, Lorsch) is used.[2] Details of the programming and technical aspects can be found in [19]. In Fig. 2.4 the general scheme of the electronic setup is shown. A small sinusoidal voltage $V_m \sin(\omega_m t)$ with amplitude V_m and frequency ω_m, generated by the lock-in is added to the applied bias voltage V. After the I/V-converter, the signal is split into a direct (DC) and an alternating (AC) signal using a low pass and a band pass, respectively. The DC part enters the feedback control which is carried out by a second DSP-board and the AC signal is amplified and measured by the lock-in amplifier.

2.2.6 Basic Principle

In the following, the main concept and advantages of the lock-in technique will be introduced. The basic idea is to modulate the tunneling voltage. A sketch of the time dependent voltage V is shown in Fig. 2.5a.

The voltage modulation results in an alternating current (AC) signal. The amplitude of the current oscillation is directly connected to the slope of the *I*(*V*)-curve, see Fig. 2.5b. Using the model of Tersoff and Hamann the tunneling current is given by the following formula

---

[2] An identical board takes care of the z–piezo displacement and realizes the feedback loop.

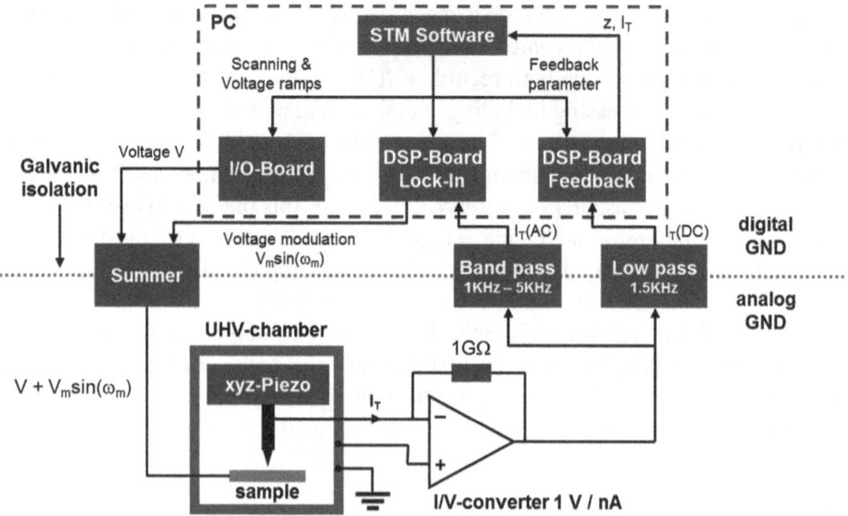

Fig. 2.4 Implementation of the digital lock-in technique into the electronic setup. The digital lock-in is realized by a digital signal processor (DSP) board. The modulation is added to the bias voltage and applied to the sample. Afterwards, the signal is split into a direct current (DC) and an alternating (AC) going to the feedback control and to the lock-in

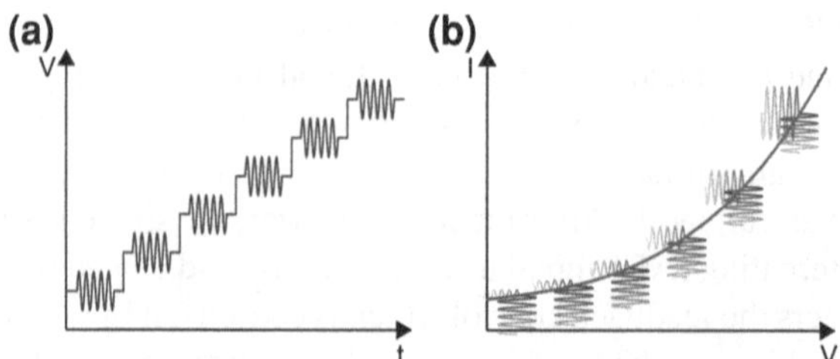

Fig. 2.5 Scheme of the lock-in technique. **a** The tunneling voltage is modulated by a sinusoidal voltage. **b** The voltage modulation leads to an alternating current signal. The amplitude of the alternating current can be connected to the first derivative of the *I*(*V*)-curve

$$I \propto \int_{0}^{eV+V_m\sin(\omega_m t)} \rho^S(\varepsilon)d\varepsilon \tag{2.11}$$

Expanding the current in a Taylor series (around $t = 0$) yields

$$I(t) \propto \sum_{n=0}^{\infty} \frac{(V_m)^n}{n!} I^{(n)}(V)\sin^n(\omega_m t) \tag{2.12}$$

where $I^{(n)}(V)$ is an abbreviation for the n-th derivative of the current I at the voltage V. By using an algebraic identity for the sine function one obtains directly the Fourier series of the current.

$$I(t) \propto \sum_{n=0}^{\infty} \frac{(V_m)^n}{n!} I^{(n)}(V) \sum_{k=0}^{n} \frac{1}{2^n} \binom{n}{k} \cos\left[(n-2k)\left(\omega_m t - \frac{\pi}{2}\right)\right]$$
$$= \sum_{n=0}^{\infty} A_n \sin(n\omega_m t + \phi_n) \tag{2.13}$$

Typically, the Fourier spectrum of the current will have only a few harmonics of non-negligible amplitude A_n. The direct current part, the A_0 term, has normally the highest magnitude of all frequencies. For the alternating current ($A_{n>0}$) usually only the amplitude A_1 of the fundamental frequency ω_m is significant. The amplitude of all higher frequencies is comparatively small. This approximation is justified in cases where the $I(V)$-curve nearly shows a linear behavior in the voltage range $[V - V_m; V + V_m]$.

The amplitude of the modulation frequency A_1 has various contributions. Explicitly, up to the fifth order, these terms are

$$A_1 = V_m I^{(1)}(V) + \frac{1}{8}(V_m)^3 I^{(3)}(V) + \frac{1}{96}(V_m)^5 I^{(5)}(V) + \cdots \tag{2.14}$$

Within the same argument as before, the linear dependency of the $I(V)$-curve in the considered voltage interval, the amplitude A_1 of the current modulation with frequency ω_m is proportional to the first derivative of the $I(V)$-curve and therefore to the LDOS of the sample.

The first derivative of the $I(V)$-curve can then be calculated by the normalized integral over time $\tau \gg 2\pi\omega_m^{-1}$ of the alternating current signal multiplied by a reference signal

$$\frac{dI}{dV}(V) = \frac{2}{V_m \tau} \int_0^{\tau} \underbrace{\sin(\omega_m t + \varphi)}_{Reference} I(t)dt \tag{2.15}$$

The reference signal is taken directly from the modulation generator and is phase shifted by φ. This phase shift is important since the alternating current signal is not in phase with the voltage modulation due to electronic components and filters. Another important aspect is the capacitance C resulting from the connecting wires and the tip-sample geometry. This capacitance gives rise to an alternating current I_c which is independent of the tunneling current I_t. Due to the capacitance the current is phase shifted by 90° compared to the tunneling current which behaves like an ohmic resistor. The amplitude of this parasitic current can be of the same order of magnitude or even higher than the signal of interest. For example a capacitance of $C = 0.8$ pF and a modulation frequency of $f_m = 2{,}000$ Hz lead to an impedance of $X_c = (2\pi f_m C)^{-1} = 100\,\mathrm{M\Omega}$. Using a voltage modulation of $V_m = 1\,\mathrm{mV}$ this results in a parasitic alternating current $I_c = V_m X_c^{-1} = 10\,\mathrm{pA}$. Considering a tunnel resistance of $R_T = 1\,\mathrm{G\Omega}$ and the same voltage modulation V_m the signal of interest $V_m R_T^{-1} = 1\,\mathrm{pA}$ is one order of magnitude smaller than I_c. In order to obtain spectroscopic data of good quality under such conditions, the phase of the reference signal has to be adjusted correctly. To achieve this goal in the experiment the tip is retracted ($R_T = \infty$) and the phase of the reference is adjusted that the above integral in Eq. (2.15) vanishes.

So far an ideal system without noise is considered. In reality the noise determines the averaging time τ in Eq. (2.15) to obtain the intended quality. For long averaging times $\tau\omega_m/2\pi \to \infty$ only the input signal from the tunneling resistance survives. All other signatures which are not phase correlated or have a different frequency than the reference signal vanish. Considering only white noise the indirect numerical derivative and the direct lock-in signal have the same signal to noise ratio for same bandwidth and averaging times. In the case of a tunneling junction it turns out that the noise is not white but rather shows a 1/f characteristic (Schottky noise) [20, 21] and contains also system specific parasitic signals (e.g. stemming from vibrational modes). The advantage of the lock-in amplifier is to measure the first derivative directly and to shift the signal from zero frequency (as it is the case for the numerical derivative) towards the modulation frequency f_m. Using a modulation frequency which has a low noise level, therefore allows increasing the signal to noise ratio. So far the signal quality was discussed only for nearly infinite averaging times τ which are not realized in any experiment. If the averaging time τ is finite, the signal has not only contributions of the reference frequency but also components with slightly higher or lower frequencies. Since the noise of the tunnel junction shows 1/f behavior, the signal to noise ratio of a lock-in is still better compared to the numerical derivation which acquires the data around zero frequency.

2.2.7 Data Processing

For a quantitative comparison of the experimental data with theory one has to take care of the STM calibration. Inaccuracies of the piezo constants and a small misangle of x- and y-scan axis lead to a distorted image of the crystal surface. The measured data can be calibrated by taking atomic resolved topographies as a reference. As atomic resolution was achieved at least once in every experiment presented in this thesis the information of the crystallographic directions for each sample is provided. Raw data obtained on a clean Cu(100) surface (see Fig. 2.6a) reveals a sheared and stretched surface unit cell. The atomic resolved data as well as the peak position of the Fourier transformed data can be used to calculate a 2×2 Matrix to calibrate the raw data. The calibrated images are oriented with the

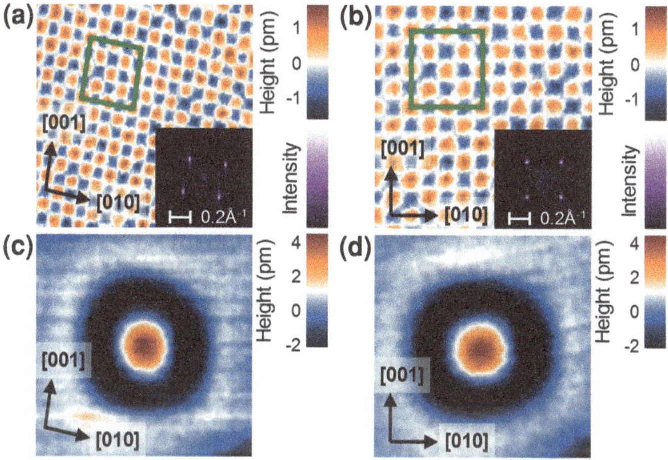

Fig. 2.6 Calibration of the measured STM data. **a** Atomically resolved topography (3 nm × 3 nm, −10 mV and 1 nA) of a clean Cu(100) surface. The corresponding powerspectrum is depicted in the *inset*. A 2 × 2 supercell of the obeserved surface unit cell is highlighted in *green*. Both the atomic corrugation as well as the peaks in the powerspectrum can be used to calibrate the measurement. **b** Calibrated topography (2.2 nm × 2.2 nm) and powerspectrum clearly showing the cubic surface unit cell of Cu(100). Again a 2 × 2 supercell is highlighted in *green*. **c** Standing-wave interference pattern of an embedded Fe impurity (2.0 nm × 2.0 nm, 10 mV and 1 nA). **d** Calibrated topography of the Fe atom (1.8 nm × 1.8 nm)

x-axis parallel to the [010] crystallographic direction. The successful data processing, indicating a cubic surface unit cell, is shown in Fig. 2.6b.

Since the sample is a single crystal, the crystallographic directions are equal for all lateral positions on the sample surface. The same shear and stretch matrix can be applied also to non-atomically resolved topographies (see Fig. 2.6c, d). The above described procedure has been applied to all data sets presented in this thesis. For most of the data sets the x- and y-axis of the topography are aligned to the [010] and [001] crystallographic direction, respectively. For the case that the crystal direction will point in other directions the coordinate system is given. When no explicit coordinate system is given in a topography the above convention for the x- and y-axis holds on.

2.2.8 STS Normalization

In order to gain and compare spectroscopic information from a *dI*/*dV*-map one has to take the tip-sample distance into consideration. Since the electronic properties of the sample are locally changed due to the presence of an impurity the tip-sample distance might not be the same for all lateral positions (x, y). If the tip is stabilized in constant current mode (topography mode), the tip height is adjusted to obtain the set-point current I_{SP} at a chosen bias voltage V_{SP}. Moreover, if thermo voltages are neglected, the current I has to vanish at $V = 0$. These two boundary conditions $I(V_{SP}, x, y) = I_{SP}$ and $I(V = 0, x, y) = 0$ restrict the variation of the dI/dV signal to the interval $V \in [V_{SP}, 0]$. For the inverse set point voltage $V = -V_{SP}$ no boundary condition exists. Therefore, the *dI/dV*-curve shows particularly strong variations on this side ($V \in [0, -V_{SP}]$). The missing information on the side of the set point voltage V_{SP} is compensated by the tip-sample distance z. This artifact can be successfully removed by normalizing the spectroscopy data to a constant tip-sample distance z^*. This method, the so-called topographic normalization, was first proposed by Garleff et al. in 2004 [22].[3]

The topographic normalization rests upon the assumption that the tunnel current depends exponentially on the tip-sample distance. The normalized spectrum can be directly obtained using the Hamer's model

$$\frac{dI}{dV}(V, x, y)\bigg|_{z=z^*} = \frac{dI}{dV}(V, x, y) \cdot \exp(2\kappa[z(x, y) - z^*]) \tag{2.16}$$

where κ is defined by the following relation

$$\kappa = \sqrt{\frac{2m_e\Phi}{\hbar^2}} \tag{2.17}$$

Here $\Phi = \left(\Phi^T + \Phi^S\right)/2$ is the mean value of tip Φ^T and sample Φ^S work function. For the calculation of κ it is assumed that the energy dependence of the transmission coefficient can be neglected for small energies and the work function Φ does not depend on the lateral position of the tip. As reference distance z^* usually the tip-sample distance of the free surface far away from an impurity is used. Due to the topographic normalization spectra of different lateral positions are comparable.

In this work a mean value of tip Φ^T and sample Φ^S work function of $\Phi = 4.2\,\text{eV}$ is used. This value is in good agreement with literature values for the work function of the pure Cu(100) surface ($\Phi^S = 4.59\,\text{eV}$) and of polycrystalline tungsten ($\Phi^T = 4.55\,\text{eV}$) [24]. Moreover, measurements of the tip-sample distance as function of the set point current give a similar value for the mean barrier height ($\Phi = 4.27\,\text{eV}$) [6].

[3] An identical formalism was introduced before by Feenstra [23]. In this paper he considered a tip movement towards the surface while acquiring STS data at *one* lateral tip position.

References

1. Quaas N (2003) Scanning tunnelling microscopy of co-impuried noble metal surfaces: Kondo-effect, electronic surface states and diffusional atom transport. Ph. D. thesis, Georg-August-Universität zu Göttingen
2. Weismann A (2003) Rastertunnelspektroskopie auf magnetisch verunreinigten Cu- und Ag-Einkristalloberflächen. Diploma thesis, Georg-August-Universität zu Göttingen
3. McGuire GE (1979) Auger electron spectroscopy reference manual. Plenum, New York
4. Weismann A (2008) Scanning tunneling spectroscopy on subsurface magnetic atoms in copper: electron focusing and Kondo effect. Ph. D. thesis, Georg-August-Universität zu Göttingen
5. Kloth P (2010) Quantentopfzustände in metallischen Heterostrukturen. Diploma thesis, Georg-August-Universität zu Göttingen
6. Prüser H (2008) Rastertunnelspektroskopie an magnetischen Atomen: Kondo-Effekt von Eisen und Kobalt in Kupfer. Diploma thesis, Georg-August-Universität zu Göttingen
7. Binnig G, Rohrer H, Gerber C, Weibel E (1982) Phys Rev Lett 49:57
8. Binnig G, Rohrer H, Gerber C, Weibel E (1983) Phys Rev Lett 50:120
9. Bonnell D (2001) Scanning probe microscopy and spectroscopy, 2nd edn. Wiley-VCH, New York
10. Chen CJ (1993) Introduction to scanning tunneling microscopy. Oxford University, USA
11. Blügel S (2010) Theorie der Rastertunnelmikroskopie. Script
12. Besocke K (1987) Surf Sci 181:145
13. Druga T (2014) Graphen auf Siliziumcarbid: elektronische Eigenschaften und Ladungstransport. Ph. D. thesis, Georg-August-Universität zu Göttingen
14. Bardeen J (1961) Phys Rev Lett 6:57
15. Tersoff J, Hamann DR (1983) Phys Rev Lett 50:1998
16. Tersoff J, Hamann DR (1985) Phys Rev B 31:805
17. Hamers RJ (1989) Annu Rev Phys Chem 40:531
18. Wahl P, Diekhöner L, Schneider MA, Kern K (2008) Rev Sci Instrum 79:043104
19. Eggebrecht T (2010) Entwicklung eines Lock-In-Verfahrens zur hochauflösenden Rastertunnelspektroskopie. Bachelor thesis, Georg-August-Universität zu Göttingen
20. Möller R, Esslinger A, Koslowski B (1990) J Vac Sci Tech A: Vac Surf Films 8:590
21. Koslowski B (1996) AIP Conf Proc 371:149
22. Garleff JK, Wenderoth M, Sauthoff K, Ulbrich RG, Rohlfing M (2004) Phys Rev B 70:245424
23. Feenstra RM (1994) Phys Rev B 50:4561
24. Stöcker H (2000) Taschenbuch der Physik. Harry Deutsch, Thun

Chapter 3
Investigation of Sub-surface Atoms by STM

Scanning tunneling microscopy (STM) and spectroscopy (STS) has shown to be a universal tool to study the electronic properties of surfaces. A first clue that STM is not only restricted to surface physics (atoms or nanostructures on a surface) on metallic systems but also give access to extended sub-surface defects in metals was presented by Schmidt and colleagues in 1996. They showed that large argon-filled nanocavities below a Al(111) surface create a contrast in topographic measurements [1, 2]. The spatial variation on the atomically flat surface of these embedded defects is due to the formation of quantum-well states induced between the cavity and the surface of the metal. This idea was refined by Kurnosikov and coworkers during the last years. They have investigated noble gas filled nanocavities below the Cu(100) and Cu(110) surface [3–6]. It turns out that the specific shape of the nanocavity, as well as the band structure of copper have to be taken into account to describe the interference pattern. During recent years the capability of STM to tackle even *single impurity atoms* which are buried below a copper surface has been proven by Weismann and Colleagues [7, 8]. The interference pattern on the surface caused by the sub-surface scattering centers is determined by the band structure of the underlying host material. The band structure may induce strong directional Friedel oscillations with wave vectors and amplitudes depending on the crystal direction [9]. These effects are not restricted to copper crystals but can also be observed for single argon and neon atoms embedded below a Ag(100) and a Ag(111) surface [10]. Here it turns out that the amplitude of the interference pattern strongly depends on the properties of the STM tip as they could be more sensitive either to surface or to bulk-like states.

In the first part of this chapter in Sect. 3.1 the local density of states (LDOS) caused by a single impurity in a metal will be introduced. This concept relates the interference pattern of single embedded impurities to the electronic band structure of the host material and is crucial for the understanding of the observed wave lengths and amplitudes in the STM topographies and spectroscopies. One important

H. Prüser, *Scanning Tunneling Spectroscopy of Magnetic Bulk Impurities*, Springer Theses, DOI: 10.1007/978-3-319-06385-0_3

parameter for the investigation of single sub-surface atoms is their position below the surface. The procedure to determine this value will be introduced at the end of this chapter in Sect. 3.2.

3.1 Imaging Buried Impurities

Before going into a detailed analysis of the signature induced by single sub-surface atoms and how they can be interpreted a characterization of the sample surface is presented. In Fig. 3.1 a large-scale topography (45 nm $\times$ 40 nm) of a cobalt-diluted copper layer is shown.

For iron- and silver-diluted thin copper films similar images are obtained. In general, the features can be classified for all samples (Ag, Fe and Co) by three different contrasts: First, defects below the surface (A). These ring-like features, which exhibit a height variation of a few picometers (-3 pm...3 pm). Contrasts of different lateral sizes belong to different impurity positions below the surface (see Sect. 3.2). A partially overlapping of single defect patterns indicating that two impurity atoms are very close to each other is also observed.

Another contrast is induced by clusters or single atoms situated in the first layer (B). They manifest themselves as dips down to -40 pm in the topography. These defect patterns arise from cobalt, iron or silver atoms which segregate to the surface during the preparation. The last types of defects are adatoms (C). These are mainly light atoms from the residual gas of the UHV chamber that accumulate after the preparation on the surface. In most cases they exhibit a positive peak with a height of about 20 pm or even higher amplitude in the topography. The distinction between atoms or clusters on the surface or in the first layer is not trivial which can be illustrated by hydrogen, the element that makes up the largest fraction in the residual gas. A hydrogen atom on a Cu(100) surface results in a dip in the topography [11].

Since the task of this thesis is to investigate sub-surface atoms by STM in this section the scattering behavior of individual impurities in a real crystal will be discussed. Usually a copper crystal is considered to be a simple metal, in which the electron system can be described by a free electron gas. Then, the scattering behavior of a single impurity has no preferred spatial directions. In contrast to this expectation the signatures of individual atoms below the surface in a constant current topography is not isotropic. This is illustrated in Fig. 3.2. The topography of a single Ag impurity buried below a Cu(100) surface (see Fig. 3.2a) shows a short period standing-wave pattern which has four-fold symmetry and an amplitude in the order of a few picometers. The experiment exhibit strongly enhanced intensities in certain directions and much weaker intensities for other directions. Topography cross sections along different directions (see Fig. 3.2b) reveal a wave length of $\lambda_{[001]} = 5.8\,\text{Å} \pm 0.1\,\text{Å}$ for the [001] direction and $\lambda_{[01\underline{1}]} = 5.4\,\text{Å} \pm 0.1\,\text{Å}$ for the $[01\underline{1}]$ direction.

The topography data shows that the scattering behavior of impurities in copper cannot be explained by an isotropic electron system. In the following, the

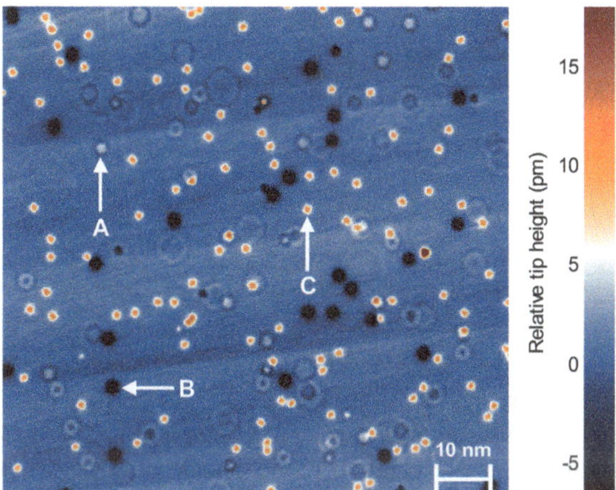

Fig. 3.1 Surface overview of a Co-diluted copper thin film sample. STM topography (45 nm × 40 nm, −10 mV, 2 nA) showing three different contrast patterns: Buried Co impurities which show only a signature height of a few picometers (**a**), Co defects assumed to be situated in the surface layer (**b**) and unknown adatoms (**c**) which are identified by their *circular shape* with a height of about 25 pm

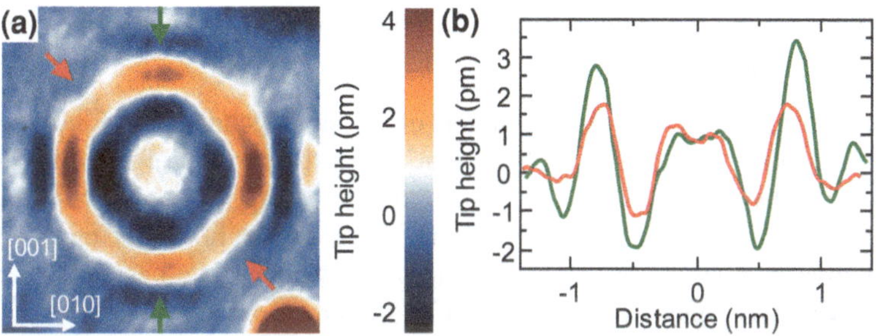

Fig. 3.2 Anisotropy of the interference pattern of a single sub-surface impurity. **a** Shows a topography (2.7 nm × 2.7 nm, −150 mV, 2 nA) of a Ag atom buried below a Cu(100) surface. **b** Corresponding cross sections along the [001] direction (*green curve*) and the [01$\underline{1}$] direction (*red curve*, marked in (**a**) by the *arrows*)

scattering behavior of an impurity in an anisotropic electron system will be discussed. In general, two quantities are of fundamental interest: First the decay of the LDOS oscillation with increasing distance and second the oscillation wave length along a certain crystallographic direction. The former quantity can be described by the electron focusing effect. A prediction of the oscillation wave length for different directions can be obtained by the stationary phase approximation. After a brief introduction to the electron focusing effect (Sect. 3.1.1) and the oscillation wave length (Sect. 3.1.2) its consequences for the Cu(100) crystal surface are discussed (Sect. 3.1.3). A more detailed description of the electron focusing model and the stationary phase approximation which are closely related, can be found in

[7, 9, 12, 13]. The concepts are generic and not restricted to copper as host material. For other metals the electron focusing should be stronger and can even causes spin density oscillation at the surface for magnetic host materials.

3.1.1 Electron Focusing Effect

Most theoretical concepts describing a perturbation potential in a metallic environment make use of the framework of Thomas-Fermi or the Friedel-screening theory to model the change of the LDOS around an impurity. These theories treat the surrounding electron system homogeneously and predict that the conduction band electrons in a metal screen the potential within a few angstroms. Calculations considering the impact of the band structure on the residual resistivity of impurities in bulk copper were first provided in the 1980s by Mertig and coworkers [14, 15]. However charge oscillations induced by the presence of buried single defects and their strange shapes were not addressed in these works.

As described above, this result contradicts the observations of buried defects in the STM (see for example Fig. 3.2). Defects up to 10 monolayer (ML) below the Cu(100) surface are observable. In addition, the impurities reveal a highly anisotropic fourfold symmetric contrast. The reason for this is the so-called electron focusing effect, which is strongly related to the electronic properties of the host material, the band structure $\varepsilon\left(\vec{k}\right)$.

In a first step, the scattering problem is discussed in general, considering a point defect at position $\vec{x}_{imp}$. The LDOS oscillations in the vicinity of the impurity can be described by the superposition of incoming and outgoing waves resulting in a standing-wave pattern. The superposition of the wave functions can be either constructive or destructive. A first indication, in which directions constructive or destructive interference occurs, is given by the group velocity:

$$\vec{v}_G\left(\vec{k},\varepsilon\right) = \nabla_{\vec{k}}\varepsilon\left(\vec{k}\right) \tag{3.1}$$

The group velocity is the surface normal of the iso-energy surface of the band structure. In areas, where the orientation of the group velocity hardly changes, and therefore the curvature of the surface is small, it comes to constructive interference of the individual wave functions, and thus to an electron focusing in the direction of $\vec{v}_G$. High positive or negative curvature indicates areas where destructive interference occurs. This has consequences for any real crystal system: The band structure reflects the anisotropic electron propagation. In analogy to a transition from a three dimensional to a one dimensional free electron system, the focused propagation and therefore reduced geometry leads to an decreased decay of the modulation in the LDOS along the focusing directions. For this reason in anisotropic media the LDOS modulations of impurities are observable at much longer distances along well-defined directions than compared to isotropic systems.

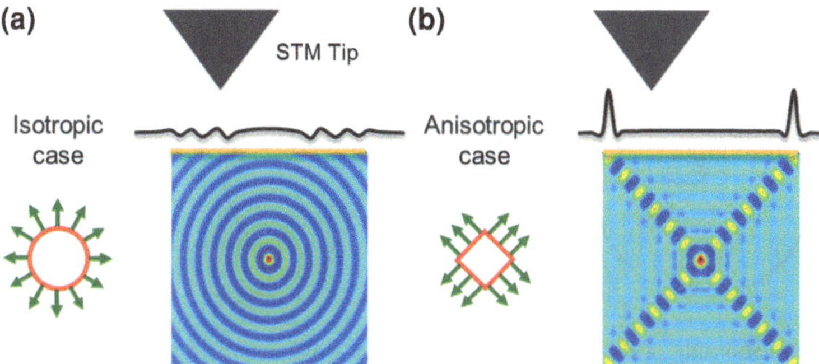

Fig. 3.3 2D model illustrating the dependence of the LDOS oscillations on the geometry of the Fermi surface. For instance two extreme cases are shown. **a** A spherical Fermi surface leads to spherical Friedel oscillations. **b** A surface composed of flat areas results in strongly anisotropic oscillations. The decay of the amplitude is much smaller for the anisotropic system (**b**) compared to the isotropic system (**a**). From [15]. Reprinted with permission from AAAS

Summing up all incoming and outgoing wave functions precisely is in general a challenging task. In Fig. 3.3 a two dimensional (2D) model is depicted demonstrating the relationship between the band structure, here the Fermi contour and the resulting LDOS oscillations at the Fermi energy along different directions. An isotropic electron system with a spherical Fermi surface leads to spherical Friedel oscillations (Fig. 3.3a). Here the decay of the amplitude is identical for all directions. An extreme anisotropic electron system, where the Fermi surface is composed of flat areas (Fig. 3.3b), results in a strongly directional dependent Friedel oscillation. The decay of the amplitude along certain directions is much smaller for the anisotropic system compared to the isotropic system.

3.1.2 Oscillation Wave Length

The electron focusing effect can explain the enhanced amplitudes of the spatial LDOS oscillation along certain directions. Another important quantity is the wave length of the Friedel oscillation for different directions. One model for the analysis of the Friedel oscillations around impurities and their wave length is the stationary phase approximation [9]. This model is valid for the far-field, this means large distances from the impurity in comparison to the oscillation wave length. For copper the Friedel wave length is in the order of 2.3 Å. The layer spacing for the Cu(100) surface is 1.8 Å, meaning that the far-field condition is fulfilled for a third-layer or deeper lying impurity. According to the stationary phase model, the oscillation periods along a certain direction $\vec{r}$ can be estimated by considering all possible group velocities $(\vec{v}_1, \vec{v}_2, \ldots)$ pointing in that direction. For the special case of only one group velocity $\vec{v}_1$ the oscillation period is determined by the projection

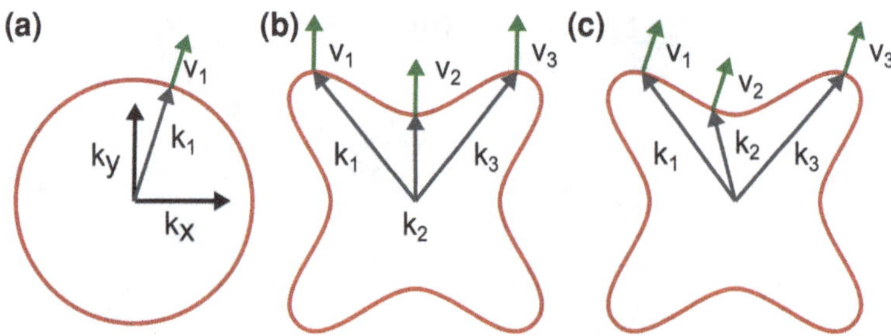

Fig. 3.4 Relationship of the spatial LDOS oscillation wave length on the Fermi contour. **a** The spherical Fermi surface has for every direction $\vec{r}$ only one point with a group velocity $\vec{v}_G$ parallel to $\vec{r}$. The artificial Fermi contour shown in (**b**) offers three points which have a group velocity pointing in the y-direction. For symmetry reasons $\vec{k}_1$ and $\vec{k}_3$ have the same period (y-component). For an inclined direction with respect to the y-axis (**c**) exist three independent points resulting in a complex oscillation for this direction

k_{r1} of the wave vector $\vec{k}_1$ onto $\vec{r}$. For two or more group velocities fulfilling the above relation, the LDOS oscillation has several periods. For example two group velocities $\vec{v}_1$ and $\vec{v}_2$ with wave vectors $\vec{k}_1$ and $\vec{k}_2$ leads to three oscillation periods $2k_{r1}$, $2k_{r2}$ and $k_{r1} + k_{r2}$. For more than two group velocities the LDOS oscillation becomes very complex consisting of many periods. In special cases, i.e. symmetry reasons, some of these periods can be equal. For instance $\vec{k}_1$ and $\vec{k}_2$ may differ but have the same projection $k_{r1} = k_{r2}$. The relationship between the Fermi surface, the group velocities and the LDOS oscillation period is illustrated in Fig. 3.4, where two artificial two dimensional Fermi contours are depicted. If for a given direction $\vec{r}$ there exist only one point on the Fermi contour with group velocity $\vec{v}_1$ parallel to $\vec{r}$, like it is the case for a spherical system, the LDOS will exhibit only one period (see Fig. 3.4a). The oscillation period is then given by two times the corresponding wave vector $\vec{k}_1$.

A more complicated Fermi contour is shown in Fig. 3.4b, c. For the direction $\vec{r}$ pointing along the y-axis, the cross-like Fermi contour offers three different points with $\vec{v}_G \parallel \vec{r}$, leading to a total of six different periods $(2k_{r1}, 2k_{r2}, 2k_{r3}, k_{r1} + k_{r2}, k_{r1} + k_{r3}, k_{r2} + k_{r3})$. However, if $\vec{r}$ is pointing along one main axis as it is shown in Fig. 3.4b, there are two points with identical projection $k_{r1} = k_{r3}$ and only three periods exist. If $\vec{r}$ points along an inclined direction, see Fig. 3.4c, there exist three independent points (k_{r1}, k_{r2}, k_{r3}). Thus for a given Fermi surface the situation can be quite complex.

3.1.3 Impact of the Surface

So far, a pure bulk system is considered. However, according to Tersoff and Hamann (see Sect. 2.2.1) the quantity of interest in an STM experiment is the

LDOS inside the vacuum at a the tip position $\vec{r}^T = (x, y, z)$, which has a certain distance z from the surface. The impact of the surface can be illustrated by a single particle approach. In general, the tunneling current has contributions of multiple states $\psi\left(\vec{k}\right)$, which contribute differently. Considering a full translational symmetry parallel to the surface, the parallel component of wave vector $k_{\parallel}$ is conserved. Inside the vacuum barrier the wave function can be described by the Schrödinger equation

$$-\frac{\hbar^2}{2m_e}\left(\kappa^2 - k_{\parallel}^2\right) = \varepsilon - \phi^{\mathrm{S}} \tag{3.2}$$

where ε is the energy of the wave function with respect to the chemical potential (Fermi energy). Using this equation a $k_{\parallel}$-dependent decay constant $\kappa\left(k_{\parallel}\right)$is found

$$\kappa\left(k_{\parallel}\right) = \frac{\sqrt{2m_e}}{\hbar}\sqrt{\left(\phi^{\mathrm{S}} - \varepsilon\right) + \frac{\hbar^2}{2m_e}k_{\parallel}^2} \tag{3.3}$$

This equation clearly demonstrates that for states with higher $k_{\parallel}$ components the effective vacuum barrier is increased and they decay faster into the vacuum. As consequence, the STM is more sensitive to states which have a small $k_{\parallel}$ value. States with higher $k_{\parallel}$ values have only a minor contribution to the tunneling current.

3.1.4 The Cu(100) Surface

In this section the before described concepts of the electron focusing effect and the stationary phase approximation are applied to the Cu(100) surface. For this purpose the band structure of copper is calculated using a linear combination of atomic orbitals approach (LCAO), which was implemented by Weismann [15]. The Fermi surface of copper and the corresponding first Brillouin zone is depicted in Fig. 3.5a. Due to the crystal potential the band structure offers no states around the L-point at the Fermi energy. This leads to eight neck-like regions along the <111> directions showing that the electron propagation is different compared to a free electron gas.

An idea of the electron propagation in copper can be obtained by looking at cross sections of the Fermi surface. In Fig. 3.5b cross sections of the [001] direction and the [011] direction (for both cross sections the second direction is the [100] direction) are shown. Looking first at group velocities pointing in the [100] direction, one immediately sees deviations to an isotropic system. One point, fulfilling $v_G \parallel [100]$ is located at the center ($k_{1\perp} = 1.42\,\text{Å}^{-1}$ with no parallel component $k_{1\parallel} = 0$). In contrast to the free electron gas (see Fig. 3.4a) there exist four

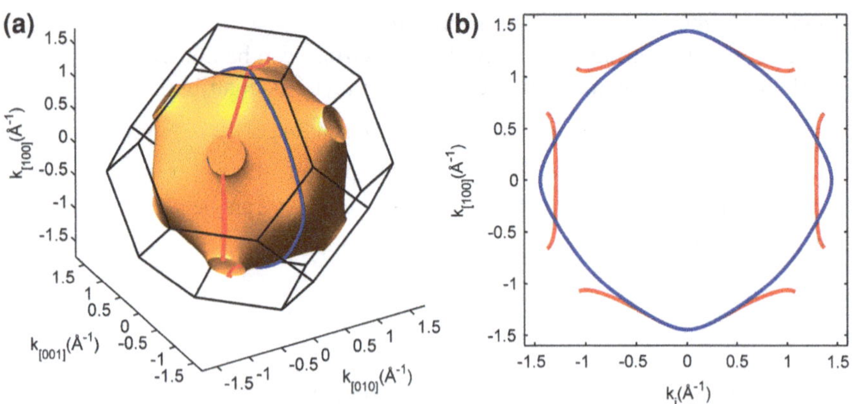

Fig. 3.5 Fermi surface of copper. **a** Shows a three dimensional representation of the Fermi surface. Cross sections (highlighted in **a**) for the $i = [001]$ (*blue*) and $i = [011]$ (*red*) direction are depicted in (**b**)

other points at the neck regions. Due to symmetry they have an identical projection $k_{2\perp} = 1.06\,\text{Å}^{-1}$ and a parallel component $k_{2\parallel} = 1.0\,\text{Å}^{-1}$. According to the stationary phase model these two independent points of the Fermi surface result in a charge oscillation along the [100] direction with periods $2k_{\perp 1} = 2.84\,\text{Å}^{-1}$, $2k_{\perp 2} = 2.12\,\text{Å}^{-1}$ and $k_{\perp 1} + k_{\perp 2} = 2.48\,\text{Å}^{-1}$. However the parallel component of the second point $k_{2\parallel} = 1.0\,\text{Å}^{-1}$ at the neck region results in a higher effective vacuum barrier. Following Eq. (3.3) the parallel component leads to an effective barrier height which is increased by 3.8 eV. Compared to the work function of Cu, which is of the same order of 4.0 eV, this means that the effective barrier for the peripheral point k_2 is nearly twice as high than for the central point k_1. Therefore, the STM is more sensitive towards electrons at the central point k_1.

A qualitative idea of the electron focusing directions can be obtained by considering the group velocities of selected cross sections. In Fig. 3.6a the cross section for the [011] direction is depicted again but now with corresponding group velocities (green arrows). Between the neck and the top region areas with reduced curvature exist (see Fig. 3.6a), leading to an electron focusing. By looking at the group velocities an accumulation of arrows slightly canted to the [100] direction can be seen. The angle histogram of the group velocities (see Fig. 3.6b) reveals a focusing direction which has an angle of 30° with respect to the [100] direction. For most of the points involved in the focusing, the wave vector and the group velocity is not collinear, leading to a reduction of the oscillation period wave vector. Picking out one representative point $k_* = \left(0.33\,\text{Å}^{-1}, 1.33\,\text{Å}^{-1}\right)$ on the contour (grey arrow) and calculating the projection on the focusing direction yields an oscillation period $2k_*^{red} = 2.6\,\text{Å}^{-1}$. This value is slightly smaller compared to the [100] direction.

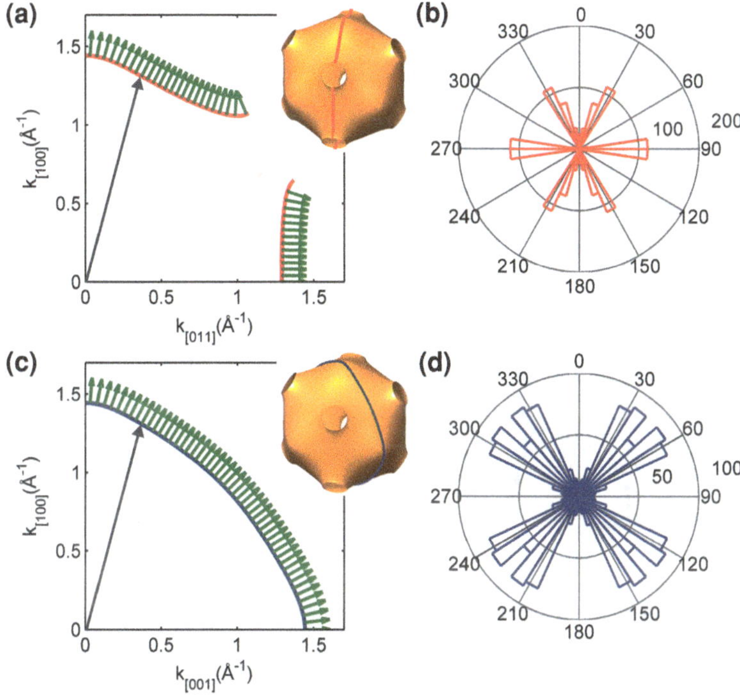

Fig. 3.6 Cross sections of the Fermi surface and group velocity of copper. **a** Cross section with the [011] plain and corresponding group velocity (*green arrows*). **b** Angle histogram of the group velocity showing an accumulation of velocities with an angle of 30° and 90°. In **c** and **d** the same analysis is depicted for the cross section with the [001] plain. Here an accumulation at 30° and 60° is observed

Doing the same analysis for the [001] cross section of the Fermi surface (see Fig. 3.6c, d) the electron focusing shows a quite similar behavior. Here also an electron focusing direction of 30° is found with nearly the same oscillation period of about 2.6 $Å^{-1}$. But compared with the [011] cross section there exists a second focusing direction with an angle of 60° with respect to the [100] direction. Since the parallel component of this second focusing direction is nearly two times higher, the STM should be more sensitive to the focusing direction with the angle of 30°. The analysis of the LDOS presented here for electrons at the Fermi energy could also be carried out for other energies. In appendix A.1 cross sections of the iso-energy surface are shown indicating that the LDOS oscillation may be considered as nearly energy independent within an energy interval of [−0.1 eV...0.1 eV].

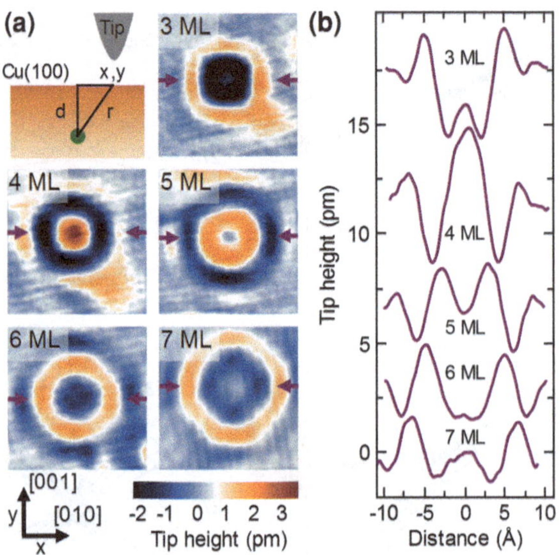

Fig. 3.7 Extracting the impurity position below the Cu(100) surface. **a** Scheme of the experiment: the *green circle* marks the impurity position below the surface and constant current topographies (2 nm × 2 nm, V = 10 mV, I = 2 nA) of single Co atoms buried below a Cu(100) surface. **b** Corresponding cross sections (shifted *vertically* for clarity) of each layer along the x = [010] direction, marked by the arrows in (**a**)

3.2 Extracting the Impurity Position Below the Surface

The embedded atoms are identified by their short period standing-wave pattern, which has fourfold symmetry (see Fig. 3.7a). The modulation in the topographic images has a wavelength of $5.5\,\text{Å} \pm 0.4\,\text{Å}$ along the [010] direction and an amplitude of a few picometers (see Fig. 3.7a).

The depth of each impurity below the Cu(100) surface is determined as follows: All standing-wave patterns on the surface are ordered to their lateral size. According to the electron propagation in copper the depth of the impurity is a monotone increasing function of the size of the standing-wave pattern.

As a verification of the found impurity position atomic resolved topographies, which exists for most of the data presented in this work, are considered (for example see Fig. 3.8a). Since the Cu(100) has a fcc-crystal structure with "ABAB" stacking and the impurity atoms are situated on substitutional lattice sites, the impurity contrast has to follow a certain ordering with respect to the host lattice. By analyzing the position of the interference pattern with respect to the surface atoms one can directly determine whether the impurity is located in an even or an odd monolayer (see Fig. 3.8b). The surface layer is labeled in the following as first monolayer. Then the standing-wave pattern of an odd monolayer impurity is centered on a corrugation maximum of the surface. If the impurity is

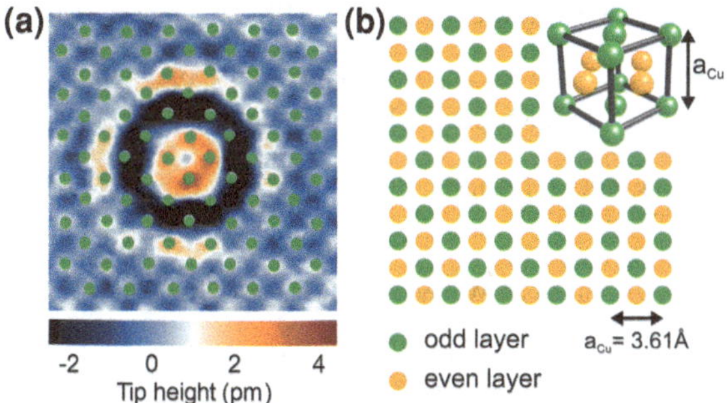

Fig. 3.8 Even or odd monolayer analysis. **a** Constant current topography (2.5 nm × 2.5 nm, −10 mV and 0.8 nA) of an Fe impurity with atomic resolution. The position of the surface atoms is marked by *green dots* for clarity. The center of the standing-wave pattern is located on an interstitial position of the surface lattice which shows that the impurity is situated in an even layer. **b** Projection of the atomic layer on the (100) surface. Two different sub lattices (marked in *green* and *orange*), belonging to even and odd monolayers are present. A three dimensional representation of the Cu fcc unit cell with marked sub lattices is illustrated in the inset

positioned in an even monolayer, the center is located between the corrugation maxima. This even or odd monolayer analysis indicates that the interference pattern shown in Fig. 3.8a belongs to an impurity atom located in an even monolayer.

The ordering by the atomic resolved data rests upon the assumption that maxima in an atomic resolved topography reflect the position of the surface atoms. This assumption is not necessarily correct, as the scanning tunneling microscope measures an electronic structure. Density functional theory simulation of the Cu(100) surface however shows that the electron density is increased above lattice sites compared to interstitial sites [16].

In addition to size information and even or odd monolayer analysis there exists density functional theory (DFT) data of single Co atoms embedded below the Cu(100) surface [7, 9, 13]. Although the ab initio simulation treats the impurity atoms not in a many-body fashion, the experimental depth analysis described above is in a good agreement with the calculated data. The actual layer depth of the impurities is determined by the size of the impurity pattern, by the even or odd monolayer analysis and by a comparison with theoretical calculations. Combining all these information the exact impurity position below the surface can be obtained.

References

1. Schmid M, Hebenstreit W, Varga P, Crampin S (1996) Phys Rev Lett 76:2298
2. Schmid M, Crampin S, Varga P (2000) J Electron Spectrosc Relat Phenom 109:71
3. Kurnosikov O, Adam OAO, Swagten HJM, de Jonge WJM, Koopmans B (2008) Phys Rev B 77:125429

4. Kurnosikov O, Nietsch JH, Sicot M, Swagten HJM, Koopmans B (2009) Phys Rev Lett 102:066101
5. Kurnosikov O, Swagten HJM, Koopmans B (2011) Phys Rev Lett 106:196803
6. Kurnosikov O, Kulikov DV, Kharlamov VS, Swagten HJM, Trushin YV (2011) Phys Rev B 84:054109
7. Weismann A, Wenderoth M, Lounis S, Zahn P, Quaas N, Ulbrich RG, Dederichs PH, Blügel S (2009) Science 323:1190
8. Heinrich AJ (2009) Science 323:1178
9. Lounis S, Zahn P, Weismann A, Wenderoth M, Ulbrich RG, Mertig I, Dederichs PH, Blügel S (2011) Phys Rev B 83:035427
10. Sprodowski C, Morgenstern K (2010) Phys Rev B 82:165444
11. Lauhon LJ, Ho W (2000) Phys Rev Lett 85:4566
12. Lounis S (2007) Theory of magnetic transition metal nanoclusters on surfaces. PhD thesis, RWTH Aachen
13. Mertig I, Mrosan E, Schöpke R (1982) J Phys F: Met Phys 12:1689
14. Mertig I, Mrosan E, Zeller R, Dederichs P (1983) H Phys Stat Sol (B) 117:619
15. Weismann A (2008) Scanning tunneling spectroscopy on subsurface magnetic atoms in copper: electron focusing and Kondo effect. PhD thesis, Georg-August-Universität zu Göttingen
16. Smith JR, Gay JG, Arlinghaus FJ (1980) Phys Rev B 21:2201

Chapter 4
Kondo Physics of Single Sub-surface Atoms

The Kondo effect has regained interest due to scanning tunneling spectroscopy (STS) experiments carried out on single magnetic impurities, which were adsorbed on a noble metal surface [1, 2].[1] Despite the sub nanometer resolution capability of this local probe technique, one of the fundamental aspects of Kondo physics, its spatial extension, is still subject to discussion. Until now all STS studies on single adsorbed atoms (for a review see [3]) and molecules [4–7] have shown that the Kondo effect manifest itself as a sharp feature at zero bias voltage. Studies, which investigate the dependence of the Kondo signature on the lateral distance, showed that the amplitude rapidly vanishes within a few angstroms and the line shape of the feature nearly remains constant with increasing distance from the atom [2, 8, 9]. Only when the electron's amplitude is amplified by placing the Kondo atom within an elliptical resonator structure the Kondo signature of a Co atom can be probed at larger distances [10]. In this quantum corral experiment Manoharan et al. detected one resonance on the adatom and another resonance in the second focus of the ellipse. However, no line shape variations with distance were reported. This is in contrast to theory [11–13], which predicts a long-range visibility and oscillatory behavior of the spectral function of the itinerant electrons—the local density of states (LDOS). The missing success of observing spatially extended Kondo signatures for surface magnetic atoms and molecules is up to now not fully understood and still under discussion. One ongoing debate concerns the influence of the surface state on the Kondo effect [8, 9, 14–16].

In this chapter the capability of scanning tunneling microscopy to investigate single sub-surface Kondo impurities will be discussed. Probing *singlebulk impurities* of dilute magnetic alloys by scanning tunneling microscopy and spectroscopy open a new way to study Kondo physics in real space. Spatially dependent

[1] Later on it turned out that in reference [1] no single cerium atoms has been investigated but rather Ce clusters [3]. The observed spectroscopic signatures are not related to single-impurity Kondo physics but to clusters. Nevertheless the publication triggered a lot of research interest and studies of magnetic atoms and molecules adsorbed on noble metal surfaces.

H. Prüser, *Scanning Tunneling Spectroscopy of Magnetic Bulk Impurities*,
Springer Theses, DOI: 10.1007/978-3-319-06385-0_4

low temperature spectroscopic information is used to map the itinerant electrons—the conduction electrons at the Cu(100) surface in the vicinity of sub-surface cobalt (Co) and iron (Fe) impurities. In contrast to adatom systems a hitherto unobserved long-range Kondo signature is observable for these systems. Single iron ($4s^2 3d^6$) and cobalt ($4s^2 3d^7$) atoms belongs to the 3d transition metals and exhibit both a partially filled d-orbital. This local moment survives putting them in copper leading to well-known Kondo systems, which were studied in the past intensively by macroscopic measurements [17]. For both systems a Kondo temperature above 6 K (the temperature of the STM) is reported. Furthermore, the choice of cobalt and iron allows testing the universal character of the Kondo effect because of their different Kondo temperatures (Fe: 8–50 K and Co: 250–1000 K).

After a short introduction to Kondo physics in Sect. 4.1, an overview of the measurements of single isolated Co and Fe atoms is given in Sect. 4.2 showing that bulk impurities offers a spatially extended Kondo feature. A theoretical interpretation of the measured signatures treating the Kondo effect as a scattering resonance right at the Fermi energy is presented in Sect. 4.3. In Sect. 4.4 a way to simulate sub-surface Kondo impurities using different state-of the art methods is introduced. During this work it turned out that the often-used Fano line shapes are not the best model for the description of the STM differential conductance around zero bias. A model, which better fits to the experimental data, is introduced in Sect. 4.5. Using this analytical expression, in Sect. 4.6 microscopic properties around magnetic impurities are explored. Furthermore the properties of iron and cobalt are compared. In Sect. 4.7 a detailed analysis of the differential conductance as function of lateral distance and impurity position below the surface is presented for both impurity atoms. The line shape of the Kondo signature shows an oscillatory behavior as function of distance, which has been predicted in the past but so far was not seen in the experiment. The oscillation period for the different directions is anisotropic, which is in very good agreement with the band structure of the underlying copper crystal. In Sect. 4.8 the Kondo resonance width—the Kondo temperature—of Fe impurities is investigated as function of distance. Finally, in Sect. 4.9 the experimental results are summarized and compared to previous works. Questions which may arise and further experiments are discussed. Some parts of this chapter especially the simulation and the many-body calculations originate from collaboration with Piet E. Dargel, Robert Peters and Thomas Pruschke from the Institute for Theoretical Physics at the University of Göttingen. The main results of this chapter, the long-range visibility of observable Kondo features and the spatial dependent analysis of the line shape have been published in [18, 19].

4.1 State of the Art

For a long time it is well-known that a localized spin degree of freedom, a magnetic impurity, in a non-magnetic host metal significantly changes the behavior of the conduction band electrons of the host metal for low temperature. Below a characteristic temperature T_k, the so-called Kondo temperature, a correlated state of impurity spin and conduction electrons builds up resulting in a variety of thermodynamic anomalies. The whole phenomenon is summarized by the term Kondo effect [20].

The most prominent experimentally hallmark of a Kondo system is the resistance minimum observed at low temperature. First experiments (see for example [21]) showing this feature were carried out in the thirties of the last century. This behavior for a metal was completely unexpected since at that time two kinds of metallic systems have been known: Superconducting materials where the resistivity drops below a certain transition temperature to zero and "normal metals" which show a constant resistance for low temperature. An example that a small concentration of magnetic impurities causes a resistance minimum for copper is depicted in Fig. 4.1. The pure copper sample shows "normal metal" behavior, meaning that the resistivity monotonically decreases and becomes constant for very low temperature. Adding a small concentration of iron atoms (0.044 %) dramatically changes the electron transport properties: below a temperature of $T = 25$ K the resistivity increases. This data [22] is only one example of experiments clearly demonstrating that the observed anomalies are caused by magnetic atoms and not related to intrinsic properties of the pure metals.

The obvious correlation of the impurity spin with the surrounding electron gas—the itinerant electrons and the universal behavior[2] awoke also theoretical interest. A first theoretical approach describing the influence of a magnetic impurity in a non-magnetic host, the s-d model was proposed by C. Zener in 1951. It contains an exchange coupling J between the conduction band electrons and the localized impurity spin [23]. In 1961 P. W. Anderson found a general model for the treatment of magnetic impurities in metals: The well-known single-impurity Anderson model[3] (SIAM) describes the impurity by a localized state, which hybridizes with the conduction band electrons [25]. Furthermore, the Coulomb repulsion for a second electron entering the orbital is taken into account. The Anderson model is up to now in the focus of research because it is one of the simplest models including many-body physics. The resistance minimum finally was explained by a seminal work of J. Kondo in 1964 using s-d model and perturbation theory up to the third order in the exchange coupling J. Taking processes, where the spin of the impurity and one conduction electron is flipped similarly into account, so-called spin-flip processes, he was able to describe the resistivity minimum [26, 27]. This explanation by J. Kondo was a milestone in the research of magnetic impurities in metals and finally the effect was named after him. Although the resistance minimum could be explained by perturbation theory, this approach results in an unphysical divergence of the resistivity for $T \to 0\,\mathrm{K}$. To overcome the difficulty new many-body techniques like the numerical renormalization group (NRG) [28, 29], quantum

[2] Temperature dependent properties of various systems in the Kondo regime follow a universal function that depends only on the ratio T/T_k.

[3] For well-defined parameters the Anderson model can be transformed to the s-d model [24].

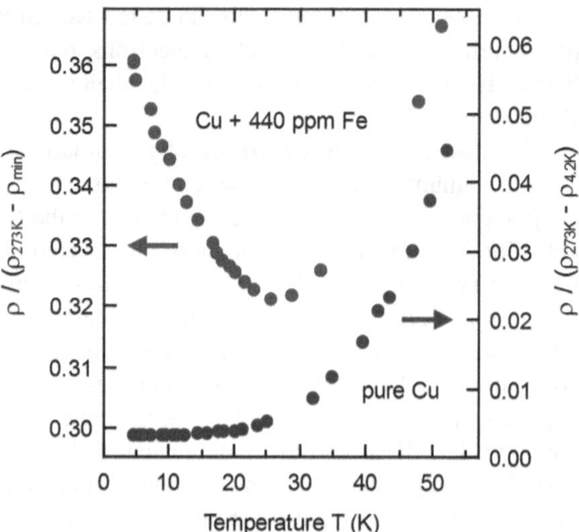

Fig. 4.1 Low temperature resistance minimum of a Kondo system. The resistance of a copper alloy containing 440 ppm iron shows a minimum as function of temperature in contrast to a pure copper sample. Reprinted after [22] with permission from Taylor & Francis Group (http://www.tandfonline.com)

Monte Carlo (QMC) [30] algorithms or the non-crossing approximation (NCA) [31] were developed.

From a microscopic point of view the most important quantity of a Kondo system is the one-particle spectrum of the impurity also called spectral function of the localized orbital. Below the Kondo temperature T_k, a narrow many-body resonance named Kondo or Abrikosov–Suhl resonance builds up in this quantity at the chemical potential leading to a singlet state of the system. The impurity spin is effectively screened by the conduction electrons. Thus the spin of both impurity and conduction electrons is correlated near the impurity and the overall spin expectation value of the whole system is zero. Since electric transport is dominated by electrons near the Fermi energy for low temperature, the Kondo resonance leads to a strong scattering of electrons at the impurity and an increase of resistivity. A sketch of the spectral function of the impurity is shown in Fig. 4.2. Above the Kondo temperature ($T \gg T_k$) two resonances one well below and one well above the Fermi energy ($E = 0\,\text{eV}$) are present. They result from the single occupied and double occupied localized orbital, respectively. Well below the Kondo temperature ($T \ll T_k$) a third resonance, the Kondo resonance appears right at the Fermi energy. The half-width at half maximum (HWHM) of the resonance $\Delta_k \sim T_k$ is proportional to the Kondo temperature.[4]

The existence of the Kondo resonance has been experimentally confirmed for dense systems with high-resolution photoemission electron spectroscopy (PES) and inverse photoemission spectroscopy (IPES) [32–36]. As a result of their

[4] According to R. Žitko and T. Pruschke the ratio between the half-width at half maximum (HWHM) and the Kondo temperature is $\Delta_k = 3.7\,k_B T_k$ [41].

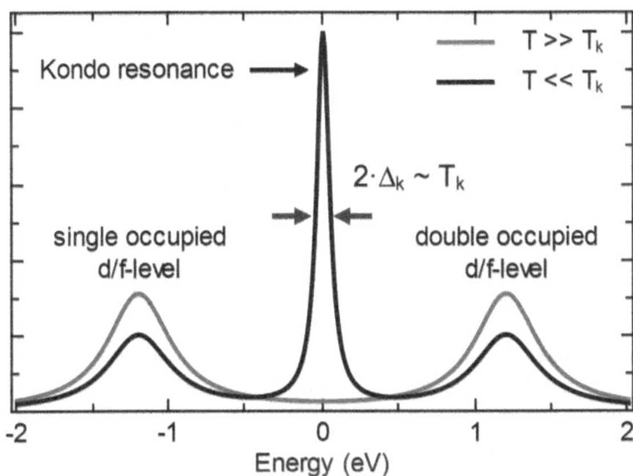

Fig. 4.2 Sketch of the spectral function of the impurity for different temperatures. For temperatures in the Kondo regime $T \ll T_k$ (*black curve*) a many-body resonance is present at the Fermi energy, which vanishes for temperatures above the Kondo temperature $T \gg T_k$(*red curve*)

limited spatial resolution these measurements as well as all macroscopic measurements always probe a very large number of magnetic atoms.

The rapidly developing field of nanotechnology paved the way to tackle single-impurity Kondo systems. In 1998 path breaking studies of scanning tunneling microscopy [1, 2] and quantum-dot devices [37, 38] were published providing information of a single Kondo impurity. Many scientists were inspired by these works and the unprecedented control over the systems. Afterwards, this time was summarized in a review article by L. Kouwenhoven [39] as "revival of the Kondo effect". The two approaches (STM and quantum-dot devices) allow studying the Kondo effect from different perspectives. Transport experiments of quantum-dot devices [40] offer a huge parameter range which can be addressed by a single device. The fundamental parameters of the Anderson model, the single electron level, the hybridization to the bath and the Coulomb interaction can be controlled in the experiment by adjusting gate voltages. The transport experiments provide local information of the impurity—the spectral function of the localized orbital, whereas the STM technique allows for the first time to study the Kondo effect in real space. It turned out that the Kondo effect can be measured by the first derivative of the current-versus-voltage $dI/dV(V)$ characteristic. In general, STM experiments cannot measure the spectral function of the localized orbital of the magnetic impurity directly, since the probability for direct tunneling into these orbitals is low due to their strongly localized character. The key quantity measured in STM experiments—in which the tip is weakly coupled to the sample—is the spectral function of the itinerant electron system, the local density of states (LDOS).

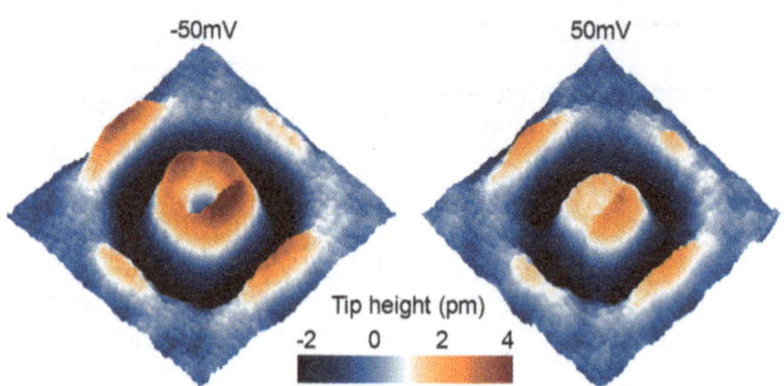

Fig. 4.3 Scattering behavior of a single sub-surface Kondo atom. STM constant current topography (1.8 nm × 1.8 nm, 2 nA) of a fourth monolayer iron impurity for different sample bias voltages. For negative bias voltage the LDOS directly above the impurity is reduced. By increasing the voltage the central ring contracts to a plateau–like maximum and the lateral extension of the interference pattern decreases

4.2 Long-Range Kondo Signature of a Single Magnetic Atom

In the previous chapter it was shown that the interference pattern caused by a sub-surface impurity is strongly influenced by the band structure of the surrounding copper crystal. The individual standing-wave patterns observed in STM topographies can be attributed to different impurity positions below the surface. In this section signatures which can be attributed to the Kondo effect of cobalt and iron impurities are of particular interest. Experimentally the fastest way to investigate the electronic structure of the sample is to measure topographies of the same area for different applied sample bias voltages while the set point current remains constant (multi-bias topographies, see Sect. 2.2.3). As a first striking example how the Kondo effect influences the energy-dependent scattering behavior near the Fermi energy, Fig. 4.3 shows two STM topographies of a fourth monolayer Fe impurity for slightly different bias voltages $V = \pm 50$ mV near zero volts.

For those small voltages the bias dependence of the transmission coefficient (Eq. 2.6) can be neglected and the topography provides directly information about the LDOS. The local minimum of the LDOS present in the center of the interference pattern for negative bias voltage ($V<0$) develops into a plateau–like maximum for positive bias voltage ($V > 0$). This change of the standing-wave pattern for this slightly different sample bias voltage cannot be explained by band structure arguments of the underlying copper crystal. The band structure of copper is nearly constant in the considered energy interval and therefore cannot cause the shift of the maxima. In the following it will be shown that the shift arises due to the Kondo effect.

Although topography measurements with different bias voltages give first insights into the scattering behavior of the impurity, they have always the drawback of averaging the LDOS over a certain energy interval. A more elaborate way to investigate the energy-dependent LDOS is by the differential conductance.

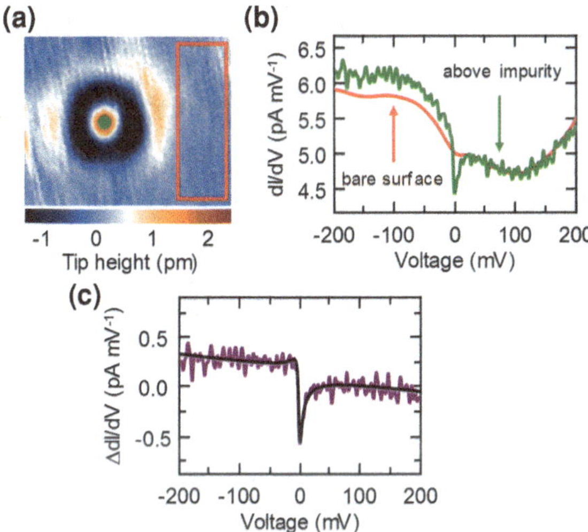

Fig. 4.4 Kondo signature in the STM differential conductance. **a** Topography of a third monolayer Fe impurity (2.5 nm × 2.0 nm, −312 mV, 1.8 nA). **b** A pair of dI/dV spectra taken with the STM tip held over the center (*green curve*) of the interference pattern, evaluated at the *green circle* and over the nearby bare Cu surface (*red curve*) evaluated above the unperturbed region of the image (*red rectangle*). **c** Difference of the differential conductance $\Delta dI/dV = dI/dV - dI_0/dV$ (*purple curve*) above the impurity (dI/dV) and above the free surface (dI_0/dV). The solid *black curve* shows a fit to the $\Delta dI/dV$ data described in the text

For instance Fig. 4.4 shows the results obtained for a third monolayer Fe atom. A typical dI/dV spectrum measured directly at the center of the interference pattern (Fig. 4.4a) reveals a strong Kondo signature, a negative peak around zero bias voltage which cannot be seen in the spectrum of the bare Cu surface (Fig. 4.4b). In order to extract the change of LDOS compared to the unperturbed surface and to remove signatures which come from the bare Cu(100) surface and the tip, the obtained differential conductance data is background corrected (see Sect. 2.2.4). The spectrum after the subtraction (Fig. 4.4c), in the following named $\Delta dI/dV$ spectrum, only shows a negative peak around zero bias, all other signatures cancelled out.

One advantage of the STM technique is that the Kondo signature induced by the impurity cannot only be investigated at one spatial position but also for other lateral tip positions. The observed signature, in particular the line shape, strongly depends on the distance to the impurity. In Fig. 4.5 $\Delta dI/dV$ spectra of a fourth monolayer Fe impurity for four different lateral tip positions A–D along the [010] direction are shown, indicating that the line shape of the spectra changes within a few angstroms. At the center of the interference pattern, direct above the impurity (position A) a positive symmetric peak is found. When increasing the lateral position the line shape becomes more asymmetric (position B), turns into a negative peak (position C) and back into an asymmetric line shape (position D). This example illustrates that a single Fe atom buried below the surface of copper

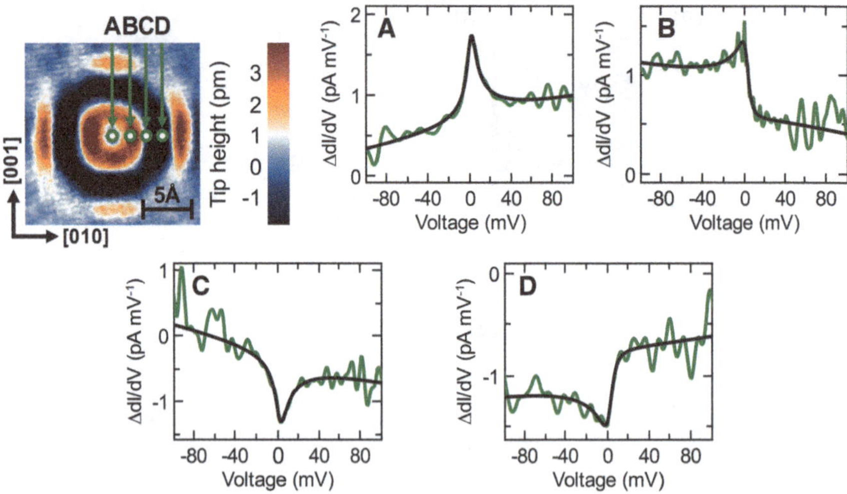

Fig. 4.5 Variation of the Kondo signature with the lateral distance of the tip. Single $\Delta dI/dV$ spectra (*green curve*) of a fourth monolayer Fe impurity for four different lateral distances along the [010] direction (A–D, marked in the topography by the *green circles*). The tip height was adjusted to give a differential conductance for the free surface dI_0/dV of 13.3 pA mV^{-1}. The *solid black curve* shows a fit to the $\Delta dI/dV$ data described in the text (see Sect. 4.5)

induces a long-range spectral signature that depends strongly on the distance to the impurity.

A detailed study of the Kondo effect and its spatial range can be obtained by the differential conductance as a function of applied sample bias voltage and one spatial coordinate, $x = [010]$ direction, across the impurity pattern, see Fig. 4.6. Single $\Delta dI/dV$ spectra correspond to vertical lines in the section (Fig. 4.6b). The local minima and maxima in the interference pattern shift to the center for positive

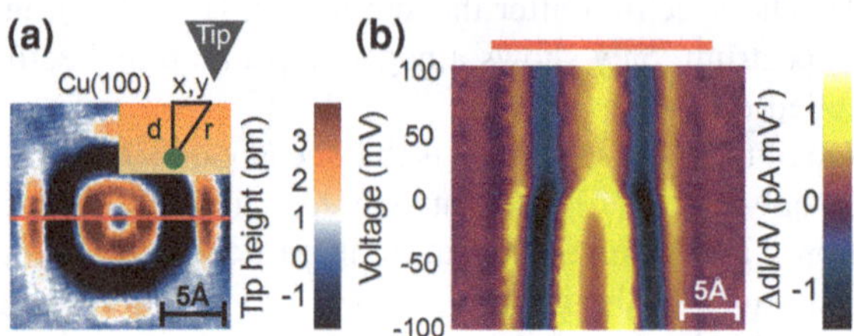

Fig. 4.6 Kondo signature of a single Fe atom as continuous function of distance. **a** Topography of a fourth monolayer Fe impurity. The scheme of the experiment is illustrated in the *inset*. The *green circle* marks the impurity below the surface. **b** $\Delta dI/dV$ signal as function of energy and one spatial coordinate, the [010] direction across the interference pattern (indicated by the red line in **a**). The tip height was adjusted to give a differential conductance for the free surface dI_0/dV of 13.3 pA mV^{-1}. An energy-dependent phase shift and enhanced amplitude of the interference pattern can be observed around zero bias voltage

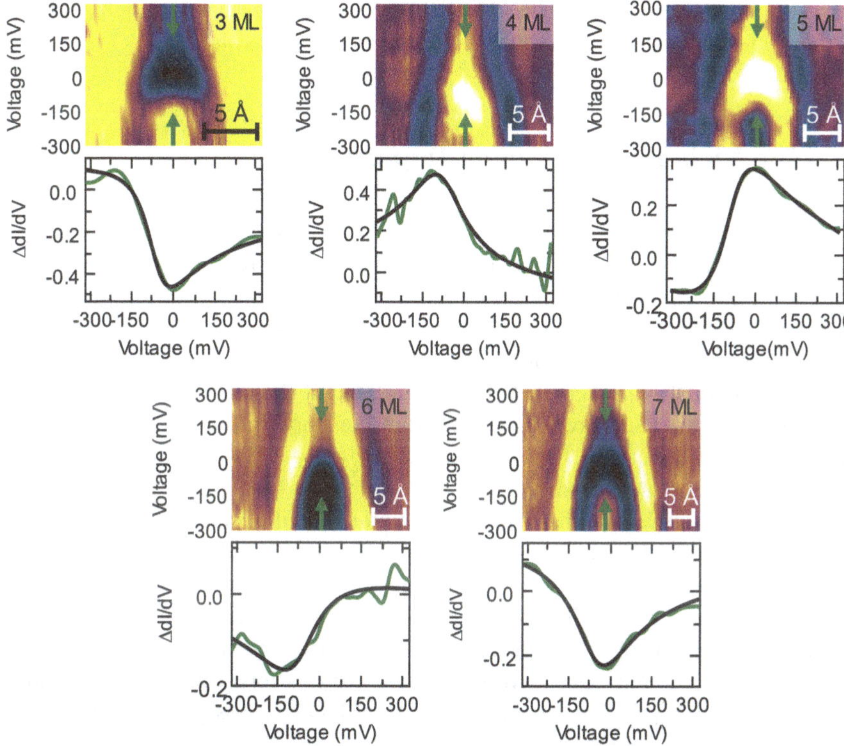

Fig. 4.7 STS data of sub-surface Co impurities in 3–7 monolayer below the Cu(100) surface. $\Delta dI/dV$ Spectrum sections along the [010] direction are shown in the *upper part*. A strong change in the scattering behavior can be seen for all impurity depths. Single $\Delta dI/dV$ spectra (*green curve*) measured with a tip position direct above the impurity (marked by the *arrows* in the sections) are depicted in the *lower part*. The tip height was adjusted to give a differential conductance for the free surface dI_0/dV of 3.2 pA mV^{-1} (3rd, 4 and 6th ML), 5.3 pA mV^{-1} (5th ML) and 6.4 pA mV^{-1} (7th ML). Note that the *color scale* for the spectroscopic data is similar to that used in Fig. 4.6b, except the range varies, and is set by the corresponding single spectra. The *solid black curve* shows a fit to the $\Delta dI/dV$ data described in the text (see Sect. 4.5)

sample bias voltages indicating a phase shift. In addition, an enhanced amplitude is observed around zero bias sample voltage.

The energy-dependent phase shift and the enhanced scattering amplitude explain the crossover in the topographies (see Fig. 4.3) and are not only found for embedded Fe impurities but also for Co atoms. An overview of the results for Co atoms situated in different depth below the surface is depicted in Fig. 4.7. The spectroscopic results are shown as single spectra (green curves) measured directly above the impurity ($x = 0$). The lateral variations of the $\Delta dI/dV$ spectra are again depicted as sections (upper part of Fig. 4.7). All impurity positions reveal an energy-dependent phase shift that results in a constriction of the interference pattern for positive bias voltages. These first results show that for sub-surface magnetic atoms a spatially extended Kondo signature is observed. In the following a model, describing the Kondo effect in the itinerant electron system, will be introduced which is able to explain the features.

4.3 The Vicinity of an Kondo Impurity

One of the most relevant quantities of a Kondo system is the spectral function of the localized orbital, which can be directly measured by high-resolution photo emission electron spectroscopy. In an STM experiment the current is determined by tunneling into the conduction electron density of states. Therefore, the STM is probing the spectral function of the itinerant electrons and not the localized ones. A first theoretical approach of the Kondo effect in the spectral function of the itinerant electrons at different distances from the impurity was given in [11–13], more recent results can be found in [42, 43].

Before presenting the main results of these works, the impact of an impurity located at position $\vec{x}_{imp}$ on the spectral function of the itinerant electrons, the local density of states LDOS at other positions $\vec{x}$ will be discussed from a general point of view. A formal description is provided by the quantum mechanical propagator: The single electron Green's function $G(\vec{x}_1, \vec{x}_2, \varepsilon)$ which describes the correlation between the creation of a single electron at position $\vec{x}_1$ and annihilation at position $\vec{x}_2$ with energy ε in real space. An introduction to this formalism can be found in [44, 45]. If the propagator is once known, the LDOS$(\vec{x}, \varepsilon)$ is given by the imaginary part of $G(\vec{x}_1, \vec{x}_2, \varepsilon)$.

$$\text{LDOS}(\vec{x}, \varepsilon) = -\frac{1}{\pi}\text{Im}G(\vec{x}, \vec{x}, \varepsilon) \tag{4.1}$$

An advantage of this formalism is the treatment of perturbations. The Green's function of an ideal host perturbed by an impurity can be derived by the Dyson equation, which connects the Green's function and hence the LDOS of the perturbed system to the Green's function $G_0(\vec{x}_1, \vec{x}_2, \varepsilon)$ of the unperturbed conduction band electrons via a T-matrix $T(\vec{x}_{imp}, \varepsilon)$. Approximating the impurity to be a point scatterer at position $\vec{x}_{imp}$ the Dyson equation reads

$$G(\vec{x}_1, \vec{x}_2, \varepsilon) = G_0(\vec{x}_1, \vec{x}_2, \varepsilon) + G_0(\vec{x}_1, \vec{x}_{imp}, \varepsilon)T(\vec{x}_{imp}, \varepsilon)G_0(\vec{x}_{imp}, \vec{x}_2, \varepsilon) \tag{4.2}$$

The T-matrix is being related to the potential perturbation and correlation effects caused by the impurity. Considering only the second part of the Dyson equation directly gives the change of LDOS compared to the unperturbed system. The resulting change includes effects arising from the band structure $G_0(\vec{x}_1, \vec{x}_2, \varepsilon)$ of the host electrons like the focusing effect and from the scattering behavior of the impurity $T(\vec{x}_{imp}, \varepsilon)$. In the following two important cases are illustrated using a simple one dimensional (1D) model system. At first, the scattering behavior of the impurity is assumed to be energy independent (see Sect. 4.3.1), which is expected to be true for non-magnetic impurities. This results in the well-known Friedel oscillation. Experimentally, silver atoms—a non-magnetic impurity—will be discussed in Chap. 5. Then, the general characteristics of a Kondo impurity, the strong energy-dependent scattering is illustrated in Sect. 4.3.2.

4.3.1 Friedel Oscillation: The Impact of the Band Structure

To demonstrate the impact of a point scatterer with energy independent scattering behavior a simple one dimensional free electron gas model can be considered. The unperturbed Green's function for this model has the following analytical form

$$G_0(\vec{x}_1, \vec{x}_2, \varepsilon) = -i\frac{m_e}{k\hbar^2}\exp(i|\vec{x}_1 - \vec{x}_2|k(\varepsilon)) \tag{4.3}$$

where the wave vector $k(\varepsilon)$ is given by the dispersion relation of the free electron gas.

$$k(\varepsilon) = \frac{\sqrt{2m_e(\varepsilon + 7\,\mathrm{eV})}}{\hbar} \tag{4.4}$$

The parameters for the free electron gas are chosen to yield an energy dependence of the wave vector where the absolute value as well as the slope is comparable with a copper crystal. The band onset is set to $\varepsilon = -7\,\mathrm{eV}$ which corresponds to a Fermi wave vector of $k_F = 1.35\ \text{Å}^{-1}$.

The complete dispersion relation of the considered model system is shown in the inset of Fig. 4.8a. In the experiment the electronic structure around the Fermi energy ($\varepsilon = 0$) is of interest. The band structure for the energy interval ($-0.1\,\mathrm{eV}\ldots0.1\,\mathrm{eV}$), highlighted in Fig. 4.8a, shows a linear behavior. Compared to the Fermi wave vector k_F, the wave vector $k(\varepsilon)$ changes less than one percent. The local density of states $\Delta\mathrm{LDOS}(x, \varepsilon)$ for a constant T-matrix ($T = -i$) is depicted in Fig. 4.8b.

The horizontal axis is the lateral distance given in multiples of the Friedel wave length $\lambda_{\mathrm{FW}} = \pi/k_F$, whereby the vertical axis is the energy. The change of the LDOS due an impurity is color coded. A localized perturbation leads to an oscillation in the local density of states. This is well-known and first predicted by J. Friedel in 1958 [46]. The run of the maxima and minima as function of energy is determined by the dispersion relation of the electron system. Since the wave vector $k(\varepsilon)$ is nearly constant for the considered energy interval, the positions of the maxima and minima are also constant. Single $\Delta\mathrm{LDOS}(x, \varepsilon)$ spectra for different distances r (Position 1–4), see Fig. 4.8c, indicates no signatures around the Fermi energy. A decreased and increased local density of states is found at position 1 ($r = 0 \cdot \lambda_{\mathrm{FW}}$) and 3 ($r = 0.5 \cdot \lambda_{\mathrm{FW}}$), respectively, while at positions 2 ($r = 0.25 \cdot \lambda_{\mathrm{FW}}$) and 4 ($r = 0.75 \cdot \lambda_{\mathrm{FW}}$) no change can be seen.

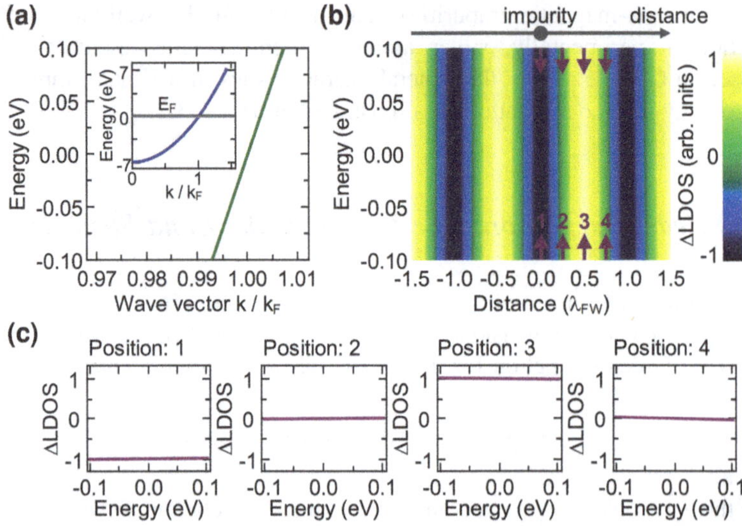

Fig. 4.8 Illustration of the local density of states in the vicinity of a point scatterer with fixed scattering amplitude and phase. **a** The dispersion relation in the considered energy interval (−0.1 eV…0.1 eV). The complete band structure is shown in the inset. **b** The ΔLDOS in the vicinity of an point scatterer shows Friedel oscillations. **c** The ΔLDOS for four different distances (1–4, in multiples of the Friedel wave length $0 \cdot \lambda_{FW}$, $0.25 \cdot \lambda_{FW}$, $0.5 \cdot \lambda_{FW}$ and $0.75 \cdot \lambda_{FW}$, marked in **b** by *purple arrows*) shows no features around the Fermi energy

It must be stressed that the main results for an energy independent point scatterer, discussed above, holds on even for two or three dimensional models and more realistic band structures. The shift of the maxima and minima is directly related to the dispersion relation. Only the spatial decay of the Friedel oscillation will differ.

4.3.2 Model of Resonant Scattering: A Kondo Impurity

A general treatment of the Kondo problem with all its details is still missing. Especially the microscopic structure of the impurity, the degeneracy of the d-, or f-orbitals, and the band structure are often neglected. However, the Kondo effect and its associated low-energy fingerprints are known to show universal behavior. Therefore, the physics in a Kondo system is to a certain extent independent of the local degrees of freedom, in particular the orbital structure of the impurity and its environment. This is important since one can apply the single-impurity Anderson model (SIAM) to interpret the experiments. Within this model the localized orbital of the impurity is described by a single level that couples to a continuum of non-interacting electrons and a Coulomb energy U between two electrons on that level. The Kondo effect in the itinerant electron system can be understand in analogy to

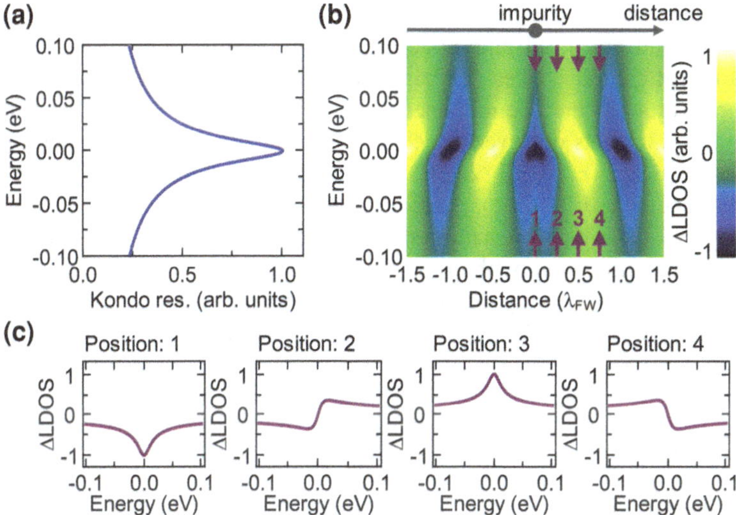

Fig. 4.9 Illustration of the Kondo effect in the itinerant electron system using a 1D model. **a** The Kondo resonance in the spectral function of the impurity causes a scattering resonance in the conduction band LDOS. The resonance width and position are fixed to $\Gamma = 10\,\text{meV}$ and $\varepsilon_k = 0\,\text{meV}$. **b** For clarity the LDOS of the unperturbed system is subtracted. The scattering resonance results in an increased amplitude of the standing-wave pattern and a phase shift when crossing the resonance. **c** For four different distances ($0 \cdot \lambda_{\text{FW}}$, $0.25 \cdot \lambda_{\text{FW}}$, $0.5 \cdot \lambda_{\text{FW}}$ and $0.75 \cdot \lambda_{\text{FW}}$, marked in **b** by *purple arrows*) the ΔLDOS is illustrated as function of energy, whereas λ_{FW} is the Friedel wave length

other fields in physics, e.g. scattering of electrons at a potential well or the "driven harmonic oscillator". The Kondo resonance leads to enhanced scattering amplitude and an energy-dependent phase shift of particles near the resonance energy. For a Kondo impurity this general characteristic can be shown by evaluating the Dyson equation with an appropriate T-matrix. For the SIAM with point like hybridization the T-matrix is just one complex function of energy and it can be shown that it is directly proportional to the spectral function of the localized orbital [11]. Here a phenomenological form found by Frota et al. is used [47, 48], which will be discussed later in Sect. 4.5.

$$T(\varepsilon) = -i\sqrt{\frac{i\Gamma}{\varepsilon - \varepsilon_k + i\Gamma}} \tag{4.5}$$

The resonance width is determined by the parameter Γ, the resonance position is given by ε_k. As a simple example, Fig. 4.9 shows the calculated change of the LDOS using the Green's function G_0 of the 1D-electron gas introduced before (see Sect. 4.3.1) and the analytical expression for the Kondo resonance (see Fig. 4.9a) described above.

The LDOS variation depicted in Fig. 4.9b clearly shows an energy-dependent phase shift and enhanced amplitude around the Fermi energy. Hence ΔLDOS spectra for different distances r (Position 1–4), see Fig. 4.9c, indicates also strong signatures around the Fermi energy. Direct at the impurity, position 1 ($r = 0 \cdot \lambda_{\mathrm{FW}}$), a dip is observed which develops into an asymmetric step-like line shape at position 2 ($r = 0.25 \cdot \lambda_{\mathrm{FW}}$). At position 3 ($r = 0.5 \cdot \lambda_{\mathrm{FW}}$) a peak is found. For a distance ($r = 0.75 \cdot \lambda_{\mathrm{FW}}$), position 4, the line shape is again asymmetric. The line shape of the Kondo signature shows an oscillation as function of distance r with a periodicity given by the Friedel wave length $\lambda_{FW} = \pi/k_F$. The prediction of the long-range visibility of the Kondo signature was already seen in the experiments (see Sect. 4.2) investigating sub-surface Fe and Co atoms.

4.4 Simulation

In this section a theoretical modeling of a sub-surface Kondo system will be discussed. The main idea is to separate the scattering behavior of the impurity affected strongly by the Kondo effect and the propagation of electrons which is given by the host material, see Eq. (4.2). The overall scheme is depicted in Fig. 4.10. A detailed description of the theoretical simulation as well as the band structure calculation can be found in [49, 50].

The free electron propagation for copper, the Green's functions $G_0(\vec{x}_1, \vec{x}_2, \varepsilon)$ can be calculated by the band structure of the material. This quantity is obtained by a linear combination of the atomic orbitals (LCAO) approach. Details of the calculation can be found in the PhD thesis of A. Weismann [49]. The resulting dispersion relation for different high symmetry directions is shown in Fig. 4.10a.

The free propagation obtained by the band structure of copper is depicted in Fig. 4.10b. Furthermore, due to the geometric setup of the experiment, the surface of the crystal plays an important role and has to be included in the simulation. This can be achieved by introducing a potential step and fixing the continuity conditions for the wave functions. In order to calculate the scattering characteristics of the impurity, the T-matrix of the system, advanced methods and models of quantum many-particle physics has to be used. The single-impurity Anderson model (SIAM) might be a good starting point modeling the Kondo effect. It is well-known that this model is appropriate to describe the universal behavior of the Kondo effect and its associated low energy fingerprints correctly. Using this model, a suitable tool calculating the T-matrix is the numerical renormalization group (NRG) [29]. At this point the LDOS—the imaginary part of the unperturbed Green's functions $G_0(\vec{x}_1, \vec{x}_2, \varepsilon)$—at the impurity position is needed to describe the states of the conduction band that couple to the impurity. Note that the presence of the surface leads to oscillations in this quantity (see Fig. 4.10c), which can be understood as a standing-wave pattern of electrons being reflected at the surface. The NRG calculations result in a Kondo resonance in the T-matrix near the Fermi energy (Fig. 4.10d), with a resonance height and width showing similar oscillatory

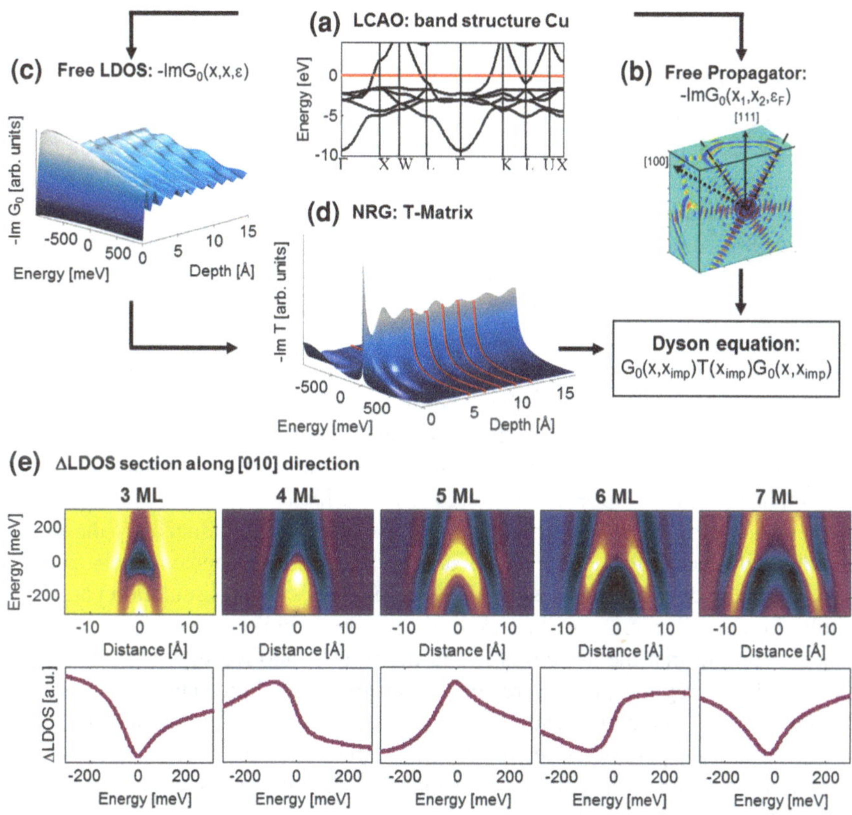

Fig. 4.10 Road map of the theoretical model. **a** Band structure of copper calculated using LCAO and **b** the related free propagator. **c** Unperturbed LDOS as function of depth, including effects of the surface. **d** The Kondo resonance for impurities located at different positions. **e** Calculated ΔLDOS section along the [010] direction for Co impurities situated 3–7 ML below the surface. Single spectra taken directly above the impurity position are shown below. Note that the *color scale* for the spectroscopic data is similar to that used in Fig. 4.6b, except the range varies, and is set by the corresponding single spectra

behavior as seen in the unperturbed Green's functions $G_0(\vec{x}_1, \vec{x}_2, \varepsilon)$. While the resonance amplitude significantly changes, the resonance width is nearly constant for different impurity positions below the surface.

With the T-matrix obtained by the above procedure, the LDOS change can be calculated using Eqs. (4.1) and (4.2). Values for the energetic position and hybridization of the localized orbital were taken from ab initio results [51]. The only free parameter left in the calculation, the Coulomb interaction U, is adjusted to yield the same resonance width as found in the experiment. The calculated LDOS sections for Co impurities, depicted in Fig. 4.10e, are in excellent agreement with the measured data (see Fig. 4.7). The experimentally observed periodicity (compare a 3rd monolayer and 7th monolayer impurity) is also found in the simulation.

Although the very good agreement of simulation and experiment seems to be convincing the model and the chosen parameter will be discussed in the following in more detail. Up to now there exist no detailed calculations for the position of the localized orbital for Co and Fe bulk impurities. The value ($\varepsilon_d = -0.8\,\mathrm{eV}$) used in the above simulation were estimated for a Co atom adsorbed on a Cu(111) surface. However this value should change going to bulk impurities and must not be the same for an Fe atom. Using the same energetic position and hybridization of the localized orbital and adjusting U for Fe impurities leads to results which describes the experiment only in the low energy regime close to the Fermi energy. For higher energies the amplitude of the T-matrix decreases leading to a vanishing signature. This behavior is not observed in the experiment and can be explained by additional scattering processes not included in the model. One additional scattering process, which naturally arises, is potential scattering due the defect potential induced by the impurity. Since the Kondo energy scale which determines the scattering behavior is normally much smaller than the energy scale induced by the defect potential, the additional process can be described by an approximately energy-independent background. In case of the Co atoms the chosen simulation parameter results in a small overlap of the broad d-resonance and the narrow Kondo resonance, meaning that the system is not in the true Kondo regime but rather in the mixed-valence regime [18]. Here the presence of the d-resonance at the Fermi energy mimic the potential scattering term actually neglected in the model. In case of Fe the chosen parameter set results in a "true" Kondo system, meaning that there is no overlap of the different resonances. The above discussion shows that potential scattering or at least an energy-independent scattering term should be added to the model. Furthermore more sophisticated values for the energetic position of the localized orbital as well as its hybridization are requisites to improve the simulation.

4.5 Phenomenological Fit Formula

As discussed in the last section, the proposed formalism, the Dyson Eq. (4.2) can be used to simulate the experimental data and to extract microscopic information of the impurities [18]. Although the simulation works quite well, all the methods used are state of the art for themselves: Combining them and finding the right parameter is a sophisticated task. To extract the key quantities from the experiment and to focus on the universal features of the Kondo effect in the low energy regime, in the following an analytical formula will be used, modeling the free propagation and the scattering matrix. This idea first proposed by Újsághy et al. [11] will be refined. As originally proposed, the free propagation of band electrons is modeled by an energy-independent phase factor $G_0(r) \approx -i\exp(ik_F r)$. The energy independence

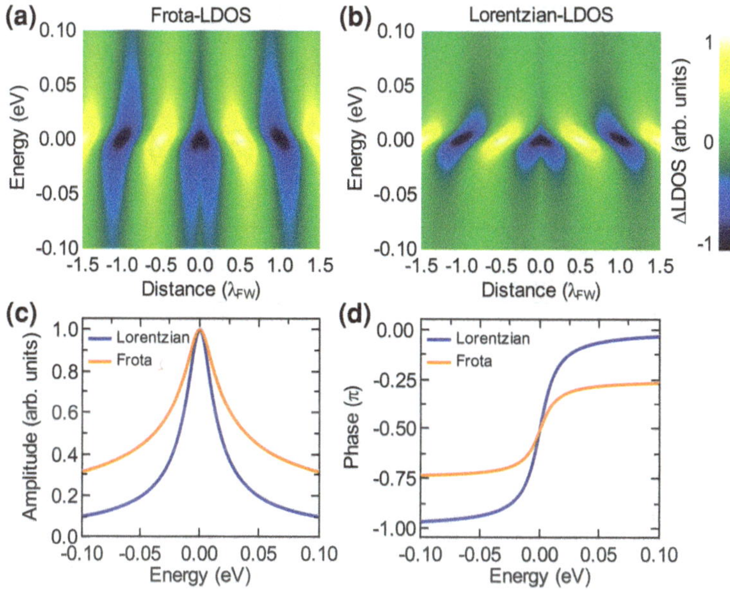

Fig. 4.11 Influence of the Kondo resonance shape on the LDOS. The change of the local density of states ΔLDOS is shown as function of distance to the impurity and energy for a Kondo resonance modeled by a Frota and a Lorentzian form, respectively, **a** and **b**. The free propagation of the electrons is modeled by a one dimensional free electron gas with a Fermi energy of $E_F = 7$ eV, see Sect. 4.3.1. The resonance width and position are fixed to $\Gamma_{\mathrm{L,F}} = 10$ meV and $\varepsilon_k = 0$ meV. **c** Scattering amplitude of the different models, both normalized to one at the Fermi energy. **d** Comparison of the phase shifts between of the Lorentzian (*blue*) and Frota (*orange*) form. While the Lorentzian shows an overall phase shift of π (starting at -1.0π going to 0.0π), the Frota form shows only a phase shift of 0.5π (starting at -0.75π going to -0.25π)

is justified as the wave vector is nearly constant around the Fermi energy $k(\varepsilon) \approx k_F$ in case of copper.

During the analysis it turned out that the widely-used (complex) Lorentzian approximation

$$T_L(\varepsilon) \propto \frac{1}{\varepsilon - \varepsilon_k + i\Gamma_{\mathrm{L}}} \tag{4.6}$$

of the T-matrix $T_L(\varepsilon)$, resembling the Kondo resonance, does not give the best description of both STM experiments [18, 52] as well as theoretical many-body calculations [18, 53]. A Lorentzian approximation of the T-matrix results in the often-used Fano line shapes [11, 54], which are depicted in Fig. 4.12. The half-width at half maximum (HWHM) $\Delta_L = \Gamma_L$ is directly given by the parameter Γ_L. Better fits were obtained by a phenomenological form $T_F(\varepsilon)$ found by Frota et al. [47, 48].

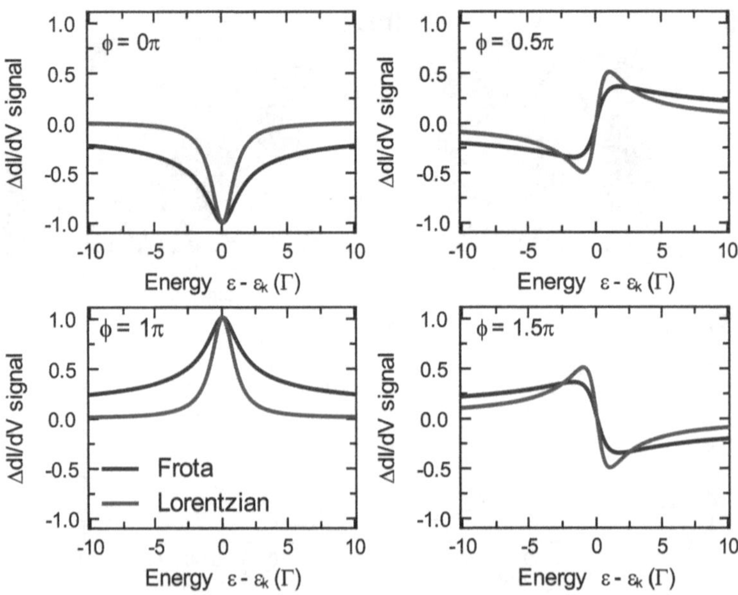

Fig. 4.12 Different line shapes of the phenomenological fit formulas. The Kondo resonance is modeled either by a Frota (*green curves*) form or a Lorentzian (*red curves*) form. The line-shape parameter ϕ determines the shape of the curve

$$T_F(\varepsilon) \propto -i\sqrt{\frac{i\Gamma_{\mathrm{F}}}{\varepsilon - \varepsilon_k + i\Gamma_{\mathrm{F}}}} \tag{4.7}$$

Here the HWHM is given by $\Delta_F = 2.54\,\Gamma_F$. In Fig. 4.11 the influence of the Kondo resonance shape to the LDOS is shown. The change of the local density of states ΔLDOS is shown as function of distance to the impurity and the energy for a Kondo resonance modeled by a Frota form (Fig. 4.11a) and a Lorentzian (Fig. 4.11b), respectively.

The scattering amplitude $|T|$ of the Frota form decays much weaker with energy than a Lorentzian, which can be viewed as a minor correction (Fig. 4.11c). More significantly, the shift of the scattering phase differs for both models (Fig. 4.11d). While the Lorentzian approximation always results in an overall phase shift of π, the Frota models yields only $\pi/2$. Recent numerical renormalization group (NRG) calculations give also a phase shift of $\pi/2$ in the maximum range of energies considered in the calculation [18]. Moreover, as one can see in Fig. 4.6b, the phase shift measured in the experiment is also smaller than π. Starting at the spatial position for an inner maximum at negative bias voltages one ends up in a knot of the interference pattern going to positive bias voltages. Assuming a phase shift of π one should end up in a minimum. This shows that the experimental phase shift is also smaller than π. Its value can be estimated to $0.4 \pm 0.3\pi$.

According to these results, the Lorentzian shape of the Kondo resonance originally used in the model of Újsághy et al. is replaced by the phenomenological form found by Frota et al. [47, 48]. The final expression for Frota's phenomenological line shape has the following functional form:

$$\frac{\Delta dI}{dV}(V) = a \cdot \mathrm{Im}\left[-ie^{i\phi}\sqrt{\frac{i\Gamma}{eV - \varepsilon_k + i\Gamma}}\right] + b \cdot V + c \tag{4.8}$$

Here ε_k is the position of the Kondo resonance. The resonance width Γ is proportional to the Kondo temperature T_k, whereas the exact relation is still an open issue. Here $\Gamma = 1.45\, k_B T_k$, is used which results from NRG calculations [41]. A linear voltage slope and an offset are added to account for additional and approximately energy-independent background scattering processes. Error bars resemble the 95 % confidence interval of the parameter and are obtained by the non-linear fit of Eq. (4.8).

The line shape of the curve is determined by the line-shape parameter (LSP) $\phi = 2k_F r$, which results for even multiples of π ($\phi = 0, 2\pi, \ldots$) in a dip and for odd multiples of π ($\phi = 1\pi, 3\pi, \ldots$) in a peak. Other values cause asymmetric line shapes. For arbitrary values of ϕ the resulting curves are plotted in Fig. 4.12 (green curve). The phenomenological fit formula, introduced above, is able to describe the experimental findings within a fairly good agreement (see Figs. 4.4 and 4.5).

4.6 Microscopic Parameters of Iron and Cobalt Atoms

In the following the phenomenological fit formula (Eq. (4.8)) is used to extract microscopic properties, the line-shape parameter, the resonance width and the resonance position of the magnetic impurities. It turns out that the difference between Fe and Co impurities can be mainly described by a different resonance width—Kondo temperature. In Fig. 4.13 cobalt and iron atoms, located both in the third monolayer, are compared. Multi-bias topographies (see Fig. 4.13a–b) demonstrate that both impurity species show similar behavior on completely different energy scales. For negative sample bias voltages a peak present in the center in of the interference pattern is found for both impurity species, which vanishes going to positive voltages. However, the voltage range the change occurs is much smaller for the Fe impurity. Here the peak is still present at $V = -3$ mV and vanishes for slightly higher bias voltage $V = 3$ mV. In contrast, for Co the topography does not change going from $V = -10$ mV to $V = 10$ mV.

More clearly the different energy scales can be seen in the $\Delta dI/dV$ spectra measured directly at the center of the interference pattern, shown in Fig. 4.13c, d. For both Co and Fe a dip-like feature is found, which can be well described by the fit formula. The line-shape parameter ϕ is nearly the same, but the resonance width

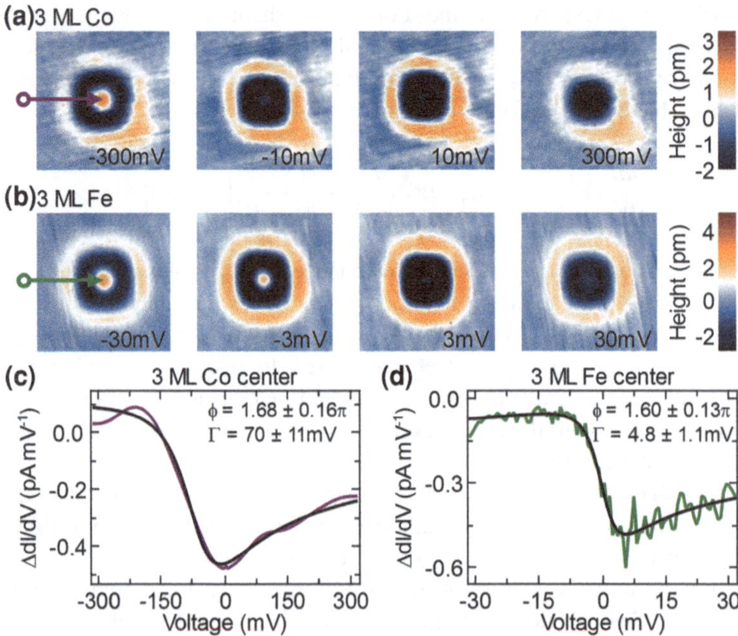

Fig. 4.13 Comparison of cobalt and iron impurities. **a** Set of constant current topographies (2 nm × 2 nm, 2 nA) for different sample bias voltages of a third monolayer Co impurity and Fe impurity (2 nm × 2 nm, 1.8 nA), **b** Topography data taken from [18]. For both impurities the peak present in the center at negative sample bias voltages vanishes for positive sample bias voltages. However, the energy scale is much smaller for the Fe atom. Single $\Delta dI/dV$ spectra measured at the center of the interference pattern, direct above the impurity (marked by the *arrows* in **a** and **b**) in the case of Co and Fe, **c** and **d**, respectively. The same line shape is observed, but the resonance width is much smaller for the Fe impurity

Γ is different. For the Fe atom a resonance width of $\Gamma = 4.8 \pm 1.1$ mV is extracted. For the Co atom a higher value of $\Gamma = 70 \pm 11$ mV is found. One question naturally arising is how the extracted parameters are related to properties of the tip. Appendix A.2 provides a detailed analysis of the Kondo signature found for different tips. It turned out that no significant changes for the extracted parameters are observed for different tips.

The spatially extension of the Kondo signature allows exploring microscopic properties around magnetic impurities. Measurements averaged over different impurity depths and spatial positions yield a resonance width $\Gamma(\text{Fe}) = 4.5 \pm 1.5$ meV and $\varepsilon_k(\text{Fe}) = 0 \pm 3$ meV for Fe impurities and $\Gamma(\text{Co}) = 81 \pm 19$ meV and $\varepsilon_k(\text{Co}) = -65 \pm 25$ meV for Co atoms. The observation that for Co the Kondo resonance is positioned well below the Fermi energy is in good agreement with recent calculations [55]. The estimated resonance width corresponds to a Kondo temperature of $T_k(\text{Fe}) = 37 \pm 12$ K and $T_k(\text{Co}) = 655 \pm 155$ K for Fe and Co atoms, respectively. These values are in good agreement with Kondo temperatures obtained from macroscopic bulk measurements (Fe: 8–50 K and Co: 250–1000 K) studied in the past [17]. However, the values of macroscopic measurements are distributed over a broad range. Also the underlying theories might differ. Therefore, the good agreement has to be interpreted with caution. Since STM has the unique capability to address only one single impurity, this might be the best way to define the Kondo temperature of a single bulk impurity.

4.7 Distance Dependence of the Kondo Signature

The idea that the line-shape parameter of the Kondo signature should depend on the distance to the impurity has been discussed theoretically in [11–13]. According to these works the line shape of the Kondo signature should show an oscillatory behavior as a function of distance. So far this theoretical prediction could not be proven for magnetic atoms or molecules adsorbed on noble metal surfaces, due to the localized visibility of the Kondo signature in these experiments. Previous sections of this chapter showed that a long-range Kondo signature is observable for single sub-surface iron and cobalt atoms. The extended spatial Kondo feature can be understood by a scattering resonance in the LDOS, the itinerant electron system. Since the resonance is energetically located at the Fermi energy, its dependence should be connected to the electronic properties of the host metal. The line-shape parameter ϕ should be a linear function of the distance r to the impurity with a slope given by two times the Fermi wave vector k_F, i.e. $\phi(r) = 2k_F r$. In the following the line-shape parameter ϕ is investigated in detail as function of lateral tip position (see Sect. 4.7.1) and as function of impurity position below the surface (see Sect. 4.7.2). A dependence on the tip-surface distance is not observed in the experiment for typical tunneling resistances in the order of $R_T =$ 0.01 GΩ...100 GΩ (see appendix A.3).

4.7.1 Lateral Dependence

At first the dependence on the lateral tip position (the [010] direction, see sketch in Fig. 3.7a), while keeping the impurity position below the surface fixed, will be analyzed. To illustrate this, the topographies of a seventh monolayer Co and fourth monolayer Fe impurity are shown in Fig. 4.14a–b, respectively.

A series of $\Delta dI/dV$ spectra taken with the STM tip held at various lateral spacing along the $x = [010]$ direction is shown in Fig. 4.14c–d. The center $(x, y = 0)$, directly above a seventh monolayer Co impurity Fig. 4.14c reveals a dip. Moving away the tip the line shape switches to a step-like feature at 3.5 Å and finally to a peak for a distance of 7.0 Å. In case of a fourth monolayer Fe atom the change of the line shape is even stronger. At the center an asymmetric peak is observed. Moving away from the center, the line shape becomes more asymmetric,

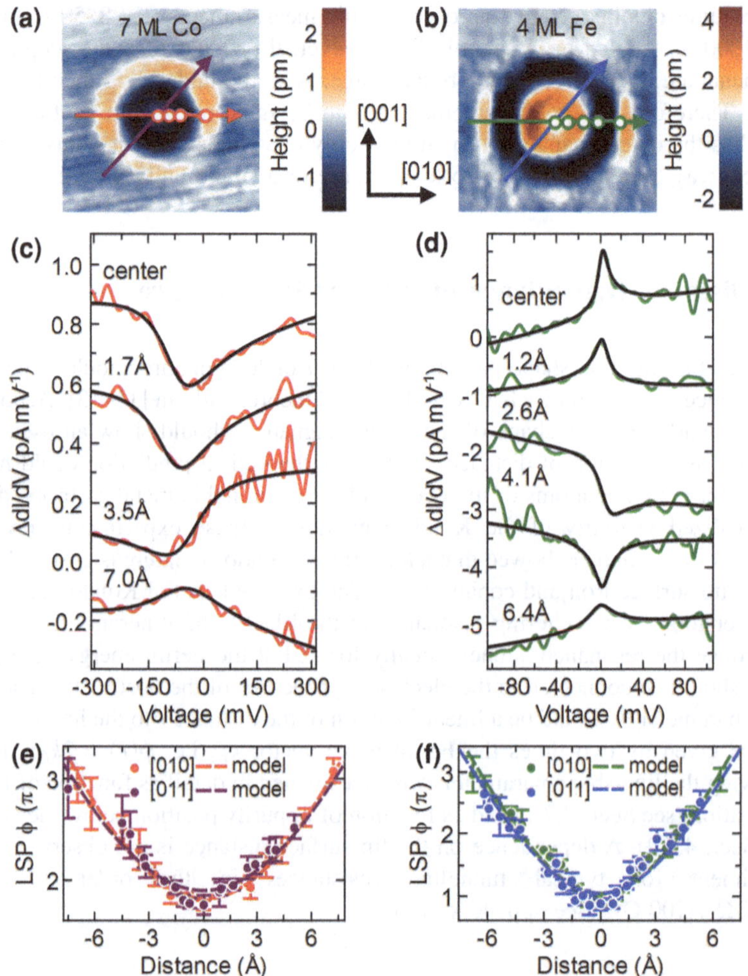

Fig. 4.14 Line-shape variation as a function of the lateral tip position. **a** STM constant current topography of a 7th monolayer Co impurity (3 nm × 3 nm, −300 mV, 2 nA) and **b** a 4th monolayer Fe impurity (2 nm × 2 nm, −150 mV, 2 nA), in which the tip position for the different spectra are depicted. Single $\Delta dI/dV$ spectra, displaced vertically for clarity, for different lateral distances, measured from the center along the [010] direction **c** for the 7th ML Co impurity and **d** for the 4th ML Fe impurity. The *solid black curve* shows fits to the $\Delta dI/dV$ data. Spatial dependence of the line-shape parameter (LSP) ϕ along the [010] and [001] direction for the Co and for the Fe impurity, respectively, **e** and **f**. The *dashed line* shows a model presented in the text

turns into a negative peak at 4.1 Å and back into a symmetric peak at a distance of 6.4 Å. In both cases, for the 7th ML Co and for the 4th ML Fe, the line-shape parameter increases continuously with the lateral tip position Fig. 4.14e–f.

As mentioned before the line-shape parameter $\phi(r) = 2k_F r$ should depend linearly on the distance r to the impurity with a slope given by two times the Fermi wave vector k_F. The distance $r = \sqrt{d^2 + x^2}$ in the experiment can be determined directly from the lateral distance x and the depth d of the impurity below the surface. This leads to the following relation $\phi = 2k_F\sqrt{d^2 + x^2} + \phi_0$ to fit the line-shape parameter. The offset ϕ_0 accounts for the line shape directly at the impurity position. For a 4th ML Fe impurity ($d = 5.4$ Å) the best fit is obtained with parameters $\phi_0 = -0.2 \pm 0.4\,\pi$ and $k_F = 0.98 \pm 0.08$ Å^{-1}. For a 7th ML Co atom ($d = 10.8$ Å) the best parameters are $\phi_0 = 0.1 \pm 0.8\,\pi$ and $k_F = 0.85 \pm 0.15$ Å^{-1}.

Band structure calculations for Cu yield values of $k_F = 1.3$ Å^{-1} in [110] and $k_F = 1.45$ Å^{-1} in [100] direction [56]. The difference between the experimental and the theoretical value is not only observed for the [010] direction but also for other crystal directions. In Fig. 4.14e–f the change of the line-shape parameter along the [011] direction is shown. Doing the same analysis introduced before yields a value of $k_F = 0.95 \pm 0.14$ Å^{-1} for a 4th ML Fe impurity and $k_F = 0.77 \pm 0.13$ Å^{-1} for a 7th ML Co impurity.

All theoretical models so far describing the line-shape dependence as a function of distance treat a homogeneous and isotropic electron gas model with a direction independent k_F. The discrepancy of the theoretical value and the wave vector found in the experiment can be basically explained by two additional ingredients: the anisotropic band structure of copper, i.e., the focusing effect and the imaging process of the STM (see Sect. 3.1). The non-collinear direction of the Fermi wave vector k_F and the group velocity v_G results in an effective reduction of the oscillation wave vector k_*, described in Sect. 3.1.1. The effective wave vector for the main focusing direction inclined by 30° to the surface normal was estimated in Sect. 3.1.3 to $k_* \approx 1.3\,\AA^{-1}$. This value is still larger as observed in the experiment. Despite the reduction caused by the band structure of the copper crystal, wave functions with different k_F contribute not equally to the tunnel current. Especially states with higher $k_{\|}$ components decay faster into the vacuum (see Sect. 3.1.2). This implies that for different lateral distances of the tip states with different k_F contribute to the tunnel current, which might indicate that the model for the lateral line-shape dependence, described above is oversimplified.

4.7.2 Dependence on the Impurity Position Below the Surface

A way to analyze the line-shape dependence in a simpler geometry is to look at $\Delta dI/dV$ spectra with different impurity depth d while the lateral tip position is kept constant at the center ($x, y = 0$) of the interference pattern, see Fig. 4.15a–e. In this geometry the contribution of tunneling paths is the same for all spectra.

The spectrum for a third monolayer Fe atom reveals an asymmetric dip. For deeper impurity positions the line shape becomes more asymmetric returning into an asymmetric dip for the seventh monolayer. The same signature is observed also for the third monolayer, indicating an oscillation length of nearly four monolayers.

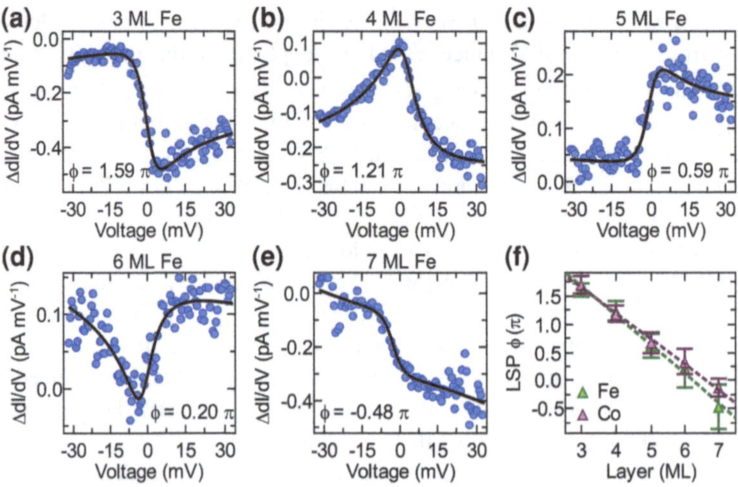

Fig. 4.15 Variation of the Kondo signature with impurity position below the surface. **a–e** Single $\Delta dI/dV$ spectra (*blue circles*) measured with a tip position direct at the center of the interference pattern for Fe atoms in monolayer 3–7. Spectroscopic measurements were performed using a lock-in technique with a modulation of the sample voltage of $V_m = 0.7$ mV at a frequency of 2,556 Hz while the tip height was adjusted to give a stabilizing current of 0.25 nA (3rd–6th ML) and 0.5 nA (7th ML) at -31 mV for the free surface. The *solid black curve* shows fits to the spectra. **f** The obtained line-shape parameter (LSP) (*green triangles* for Fe and *purple triangles* for Co) decreases linearly. The *dotted lines* show linear fits. The spectra used for the analysis of the Co atoms are depicted in Fig. 4.7

The line-shape parameter, depicted in Fig. 4.15f, apparently decreases linearly with increasing depth of the impurity. A linear fit reveals a slope of $-0.51 \pm 0.08\ \pi/\text{ML}$. This negative slope is due to aliasing between the periodicity of the lattice and the wave length of the oscillations. The line-shape parameter is 2π periodic and hence the observation is equivalent to a slope of $1.49 \pm 0.08\ \pi/\text{ML}$, resulting in a Fermi wave vector of $k_F = 1.30 \pm 0.07$ Å^{-1}. Doing the same analysis for Co impurities, see Fig. 4.15f, yields a slope of $-0.45 \pm 0.05\ \pi/\text{ML}$ leading to $k_F = 1.35 \pm 0.05$ Å^{-1}. For both impurities the obtained values are in good agreement with the expected value of copper for the [100] direction $k_F = 1.42$ Å^{-1}, see Sect. 3.1.3.

4.8 Spatial Dependence of the Kondo Temperature

In the last section the distance dependence of the line-shape parameter as function of impurity position below the surface and as function of lateral-tip position was investigated. In the following, the same analysis is performed for the Kondo resonance width—the Kondo temperature—of Fe impurities. Therefore, spectroscopic

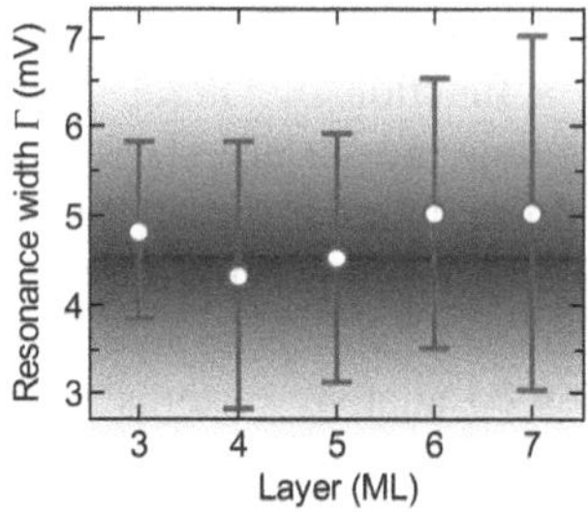

Fig. 4.16 Kondo resonance width of sub-surface Fe impurities in the 3rd–7th monolayer below the Cu(100) surface. As reference the value averaged over different impurity depth and spatial positions $\Gamma(\mathrm{Fe}) = 4.5 \pm 1.5\,\mathrm{meV}$ is highlighted by the *blue line*

measurements of different tips and impurity positions below the surface are analyzed. The results for the Kondo resonance width Γ of different impurity depth d, while the lateral tip position is kept constant at the center $(x, y = 0)$ of the interference pattern, are summarized in Fig. 4.16. For all depth a nearly constant resonance width is found, which is comparable to the averaged value over different impurity depth and spatial positions $\Gamma(\mathrm{Fe}) = 4.5 \pm 1.5$ meV.

Investigating the Kondo temperature for a constant impurity depth as function of the lateral tip position an unexpected variation of the resonance width can be observed. As an example, the fourth monolayer Fe impurity, see Fig. 4.5, shows a variation of $T_k(A) = 29 \pm 9$ K, $T_k(B) = 35 \pm 24$ K, $T_k(C) = 42 \pm 18$ K and $T_k(D) = 44 \pm 26$ K from position A to D. The rise of the Kondo temperature with distance is very small and mostly within the confidence interval of the values. Nevertheless, the trend is found for both Fe and Co and not depends on the impurity position below the surface. This behavior is not included in the single-impurity Anderson (SIAM) model. One possible explanation considers the actual orbital structure of the d-level for Fe or Co, which leads to different hybridization—screening of the spin—for the d-orbitals. Experimental findings going in the same direction were recently obtained on magnetic molecules deposited on superconducting lead [57] or gold crystal surfaces [58].

4.9 Discussion

Single magnetic Fe and Co impurities embedded below the Cu(100) surface opens a new way to study single-impurity Kondo physics by means of STM. In contrast to previous works using adatom systems, which had studied only local properties, this work focuses on single Kondo impurities in the bulk on length scales, which were so far not accessible. The long-range Kondo signature allows exploring the full microscopic properties around magnetic impurities. The spatial distribution of the Kondo signature is—in addition to the Kondo physics—strongly influenced by the anisotropy of the band structure of the host crystal. A detailed analysis of differential conductance as a function of lateral distance for Fe and Co atoms situated at different depths below the surface shows that the distance dependence of the line shape of the Kondo signature is directly connected to the band structure of the host crystal. Additionally, probing bulk impurities with STM allows separating the physics of the correlated many-body system and the measuring process. This is in contrast to studies on adatom systems. Models for tunneling through adatom or molecule systems have to treat the tip, its coupling to the localized impurity state, its coupling to the itinerant bulk electrons as well as surface states, and all interferences between alternative transport paths [12, 13]. Thus, the Kondo signature is a feature of the measurement process and is associated with the relative strength of the hybridization between the tip, sample, and impurity. In contrast, the investigation of bulk impurities allows a clear spatial separation between the Kondo impurity and STM process. Within this interpretation, the STM is a noninvasive tool that exclusively probes the LDOS of the conduction band electrons—the itinerant electrons.

Going further the long-range Kondo signature opens a new way to examine the interaction between two or more Kondo atoms with each other, or the effect of an interface in real space. First results investigating two iron atoms next to each other will be discussed in Chap. 6.

References

1. Li J, Schneider W-D, Berndt R, Delley B (1998) Phys Rev Lett 80:2893
2. Madhavan V, Chen W, Jamneala T, Crommie MF, Wingreen NS (1998) Science 280:567
3. Ternes M, Heinrich AJ, Schneider W-D (2009) J Phys: Condens Matter 21:053001
4. Zhao A, Li Q, Chen L, Xiang H, Wang W, Pan S, Wang B, Xiao X, Yang J, Hou JG, Zhu Q (2005) Science 309:1542
5. Wahl P, Diekhöner L, Wittich G, Vitali L, Schneider MA, Kern K (2005) Phys Rev Lett 95:166601
6. Iancu V, Deshpande A, Hla S-W (2006) Nano Lett 6:820
7. Iancu V, Deshpande A, Hla S-W (2006) Phys Rev Lett 97:266603
8. Knorr N, Schneider MA, Diekhöner L, Wahl P, Kern K (2002) Phys Rev Lett 88:096804
9. Henzl J, Morgenstern K (2007) Phys Rev Lett 98:266601
10. Manoharan HC, Lutz CP, Eigler DM (2000) Nature 403:512
11. Újsághy O, Kroha J, Szunyogh L, Zawadowski A (2000) Phys Rev Lett 85:2557
12. Schiller A, Hershfield S (2000) Phys Rev B 61:9036
13. Plihal M, Gadzuk JW (2001) Phys Rev B 63:085404
14. Limot L, Berndt R (2004) Appl Surf Sci 237:572
15. Schneider M, Vitali L, Wahl P, Knorr N, Diekhöner L, Wittich G, Vogelgesang M, Kern K (2005) Appl Phys A Mater Sci Process 80:937
16. Li Q, Yamazaki S, Eguchi T, Kim H, Kahng S-J, Jia JF, Xue QK, Hasegawa Y (2009) Phys Rev B 80:115431
17. Daybell MD, Steyert WA (1968) Rev Mod Phys 40:380
18. Prüser H, Wenderoth M, Dargel PE, Weismann A, Peters R, Pruschke T, Ulbrich RG (2011) Nature Phys 7:203
19. Prüser H, Wenderoth M, Weismann A, Ulbrich RG (2012) Phys Rev Lett 108:166604

20. Hewson AC (1993) The Kondo problem to heavy fermions. Cambridge University Press, Cambridge
21. De Haas W, Van Den Berg G (1936) Physica 3:440
22. Pearson W (1955) Philos Mag Ser 7(46):911
23. Zener C (1951) Phys Rev 82:403
24. Schrieffer JR, Wolff PA (1966) Phys Rev 149:491
25. Anderson PW (1961) Phys Rev 124:41
26. Kondo J (1964) Progress Theoret Phys 32:37
27. Kondo J (2006) Proc Japan Acad B 82:328
28. Wilson KG (1975) Rev Mod Phys 47:773
29. Bulla R, Costi TA, Pruschke T (2008) Rev Mod Phys 80:395
30. Pruschke T, Grewe N (1989) Zeitschrift für Physik B Condensed Matter 74:439
31. Silver RN, Gubernatis JE, Sivia DS, Jarrell M (1990) Phys Rev Lett 65:496
32. Reinert F, Hüfner S (2005) New J Phys 7:97
33. Patthey F, Delley B, Schneider WD, Baer Y (1985) Phys Rev Lett 55:1518
34. Patthey F, Schneider WD, Baer Y, Delley B (1987) Phys Rev Lett 58:2810
35. Patthey F, Imer J-M, Schneider W-D, Beck H, Baer Y, Delley B (1990) Phys Rev B 42:8864
36. Ehm D, Hüfner S, Reinert F, Kroha J, Wölfle P, Stockert O, Geibel C, v. Löhneysen (2007) H Phys Rev B 76:045117
37. Goldhaber-Gordon D, Shtrikman H, Mahalu D, Abusch-Magder D, Meirav U, Kastner MA (1998) Nature 391:156
38. Cronenwett SM, Oosterkamp TH, Kouwenhoven LP (1998) Science 281:540
39. Leo Kouwenhoven LG (2001) Phys World 14:33
40. Pustilnik M, Glazman L (2004) J Phys: Condens Matter 16:R513
41. Žitko R, Pruschke T (2009) Phys Rev B 79:085106
42. Affleck I, Borda L, Saleur H (2008) Phys Rev B 77:180404
43. Büsser CA, Martins GB, Costa Ribeiro L, Vernek E, Anda EV, Dagotto E (2010) Phys Rev B 81:045111
44. Mahan GD (2000) Many particle physics, 3rd edn. Springer, New York
45. Economou EN (2006) Green's functions in quantum physics, 3rd edn. Springer, New York
46. Friedel J (1958) Il Nuovo Cimento 7:287
47. Frota HO, Oliveira LN (1986) Phys Rev B 33:7871
48. Frota HO (1992) Phys Rev B 45:1096
49. Weismann A (2008) Scanning tunneling Spectroscopy on subsurface magnetic atoms in copper: electron focusing and Kondo Effect. PhD thesis, Georg-August-Universität zu Göttingen
50. Dargel PE (2012) Spectral functions of low-dimensional quantum systems. PhD thesis, Georg-August-Universität zu Göttingen
51. Lin C-Y, Neto AHC, Jones BA (2006) Phys Rev Lett 97:156102
52. Fu Y-S, Ji S-H, Chen X, Ma X-C, Wu R, Wang C-C, Duan W-H, Qiu X-H, Sun B, Zhang P, Jia J-F, Xue Q-K (2007) Phys Rev Lett 99:256601
53. Žitko R (2011) Phys Rev B 84:195116
54. Fano U (1961) Phys Rev 124:1866
55. Surer B, Troyer M, Werner P, Wehling TO, Läuchli AM, Wilhelm A, Lichtenstein AI (2012) Phys Rev B 85:085114
56. Segall B (1962) Phys Rev 125:109
57. Franke KJ, Schulze G, Pascual JI (2011) Science 332:940
58. Minamitani E, Tsukahara N, Matsunaka D, Kim Y, Takagi N, Kawai M (2012) Phys Rev Lett 109:086602

Chapter 5
Signatures of Non-magnetic Atoms

The last chapter was devoted to single-impurity Kondo physics of magnetic iron and cobalt atoms. In this chapter *non-magnetic* silver (Ag) impurities below the Cu(100) surface are considered. This sample system represents a control experiment to verify that the observed features for magnetic Fe and Co are really due to their localized magnetic moment. The Cu(100) surface is an excellent model system to investigate the impact of embedded silver atoms on the electronic structure of the surface and to compare the results to magnetic Fe and Co impurities. At first, in Sect. 5.1 it is shown that single non-magnetic Ag defects can strongly influence the electronic structure of the surface. Then, in Sect. 5.2 the voltage dependence of the signature will be discussed. Furthermore, a comparison of the scattering characteristics for both magnetic and non-magnetic impurities is presented. While for magnetic iron and cobalt atoms the interference pattern strongly depends on the applied sample bias voltage for silver atoms only a minor dependence is found. Nevertheless, spectroscopic maps reveal an unexpected small kink around zero sample bias voltage. In Sect. 5.3 different explanations will be discussed. Finally, the results are summarized in Sect. 5.4, followed by an outlook for further (control-) experiments.

5.1 Topographic Investigation

As a first example how the standing-wave pattern depends on the energy Fig. 5.1 shows STM topographies of a seventh monolayer Ag impurity for different sample bias voltages V. The lateral extension of the focusing pattern decreases for positive bias voltages (Fig. 5.1a, b), while the overall appearance of the standing-wave pattern does not depend on the applied bias voltage. More clearly this feature can be observed in cross sections along the [010] direction, depicted in Fig. 5.1c. The spatial positions of the maxima and minima are nearly identical for negative bias

H. Prüser, *Scanning Tunneling Spectroscopy of Magnetic Bulk Impurities*,
Springer Theses, DOI: 10.1007/978-3-319-06385-0_5

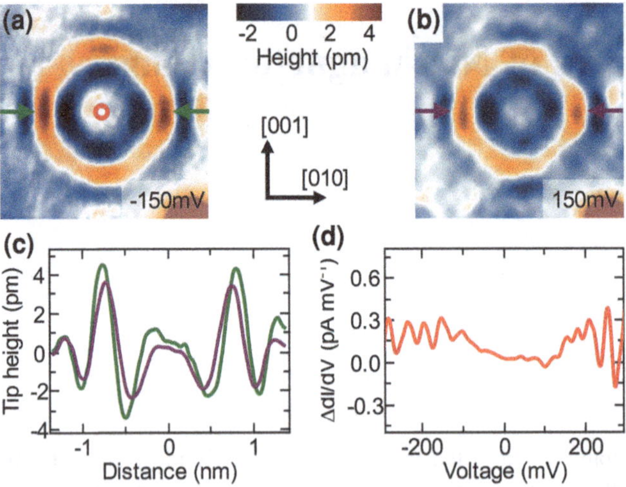

Fig. 5.1 Scattering behavior of a single sub-surface silver impurity. **a** STM constant current topography (2.7 nm × 2.7 nm, 2 nA) of a seventh monolayer silver impurity for a sample bias voltages of $V = -150\text{mV}$ and **b** $V = 150$ mV. **c** Corresponding cross sections along the [010] direction (marked in **a** and **b** by the *arrows*). For positive (*purple curve*) and negative (*green curve*) bias voltages the interference pattern hardly changes. **d** A single $\Delta dI/dV$ spectrum measured with a tip position at the *center* of the standing-wave pattern (marked by the *red dot* in **a**). The tip height is adjusted to give a set point current of $I_{SP} = 2$ nA at $V_{SP} = -\ 300$ mV for the free surface

voltage $V = -150$ mV and positive bias voltage $V = 150$ mV. This observation is in contrast to the behavior of embedded magnetic Fe and Co impurities exhibiting standing-wave patterns and spatial positions of the maxima which strongly depend on the applied sample bias voltage (see for instance Fig. 4.13).

In Fig. 5.1d a background subtracted single $\Delta dI/dV$ spectrum measured with a tip position direct above the impurity is shown. The spectrum reveals no features in the energy range considered here ($V = -300...300$ mV).

5.2 Voltage Dependence of the Interference Pattern

The energy-dependent scattering behavior of the embedded impurity can be investigated by the background subtracted differential conductance as function of applied bias voltage and one spatial coordinate across the standing-wave pattern. A $\Delta dI/dV$ section along the [010] direction for a seventh monolayer Ag impurity is depicted in Fig. 5.2.

While at the center of the interference pattern—directly above the impurity position—no signature is observed at the border a local maximum surrounded by two minima is found. A shift of the lateral position of the maximum can be observed for increasing energy within a voltage range of ±300 mV.

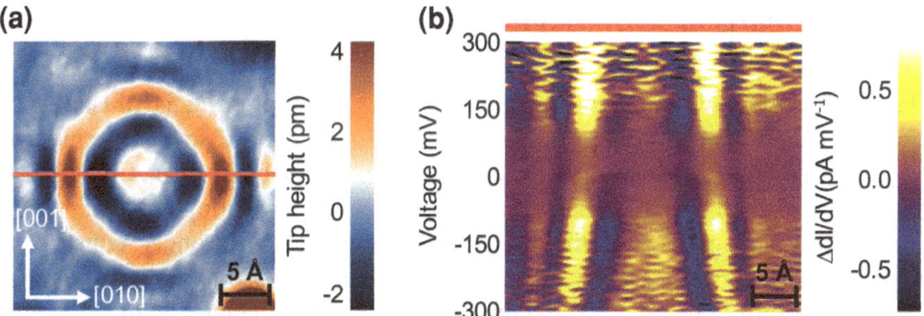

Fig. 5.2 Spectroscopic signature of a single silver atom as continuous function of distance. **a** Topography of a seventh monolayer Ag impurity (2.7 nm × 2.7 nm, −150 mV, 2 nA). **b** $\Delta dI/dV$ signal as function of applied bias voltage and one spatial coordinate, the [010] direction across the interference pattern (indicated by the *red line* in **a**). The tip height is adjusted to give a set point current of $I_{SP} = 2\text{nA}$ at $V_{SP} = -\ 300$ mV for the free surface

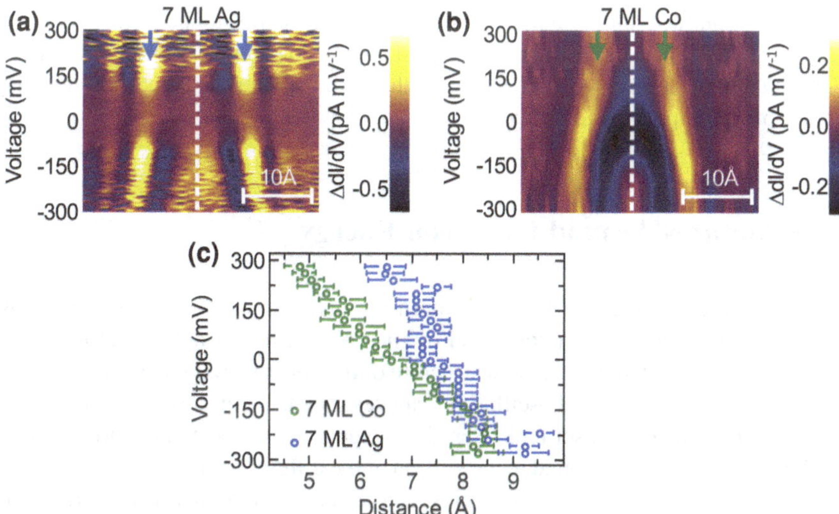

Fig. 5.3 Comparing the scattering of non-magnetic silver and magnetic cobalt impurities. **a** Section of the differential conductance along the [010] direction for a seventh monolayer Ag impurity and **b** for a seventh monolayer Co impurity. The tip height is adjusted to give a set point current for the free surface of $I_{SP} = 2$ nA at $V_{SP} = -\ 300$ mV and at $V_{SP} = -\ 312$ mV for Ag and Co, respectively. **c** The spatial position (relative to the *center*, marked in **a** and **b** by the *white line*) of the LDOS maxima as a function of applied bias voltage. The uncertainty according to the distance arises from averaging over the [010] direction and [0$\bar{1}$0] direction, as well as from data processing. In case of the Ag impurity the position of the maxima changes hardly whereas for the Co impurity the maximum strongly moves to the *center*

5.2.1 Comparison with Magnetic Impurities

The difference between magnetic and non-magnetic scattering is illustrated in Fig. 5.3. Here the section of the differential conductance of a seventh monolayer Ag impurity (Fig. 5.3a) is directly compared to the section of a Co impurity (Fig. 5.3b) located in the same layer along the same crystallographic direction. The interference pattern of the magnetic atom shows a stronger constriction increasing the voltage. More clearly this feature can be illustrated by mapping the spatial position of the $\Delta dI/dV$ maximum relative to the center of the standing-wave pattern as function of applied bias voltage V. The position of the maximum is located at 8.5-9.5 Å for a negative bias voltage of $V = -$ 280 mV and moves inwards by increasing the voltage for both, Ag and Co. However, the slopes of the curves are different. For a positive bias voltage of $V = 280$ mV the position of the maximum is 6.5 Å for Ag and 4.8 Å for Co, respectively. According to the previous chapter, the strong shift of the phase fronts for the magnetic Co impurity is related to the Kondo effect which results in a scattering resonance at the Fermi energy. In contrast, the impurity scattering of Ag seems to be nearly energy-independent for the energy range considered here. The small shift observed for Ag is caused by the change of the wavelength of the scattering electrons, the dispersion of the copper band structure.

5.3 Signatures Around the Fermi Energy

The development of the interference pattern for silver impurities embedded below the Cu(100) surface can be understood by the concept of Friedel oscillations (see Sect. 4.3.1) and the underlying band structure of the copper crystal (see Sect. 3.1.3). Although the concept Friedel oscillations can describe the experimental findings for a broad energy range, a closer look (Fig. 5.4) reveals an unexpected feature: Around zero bias voltage a kink is found in the section of the differential conductance, shown in Fig. 5.4b. The kink is located very close to the Fermi energy in a voltage range of about $V = \pm 10$ mV and exhibits only a small phase shift. This feature is not only found for a seventh monolayer Ag impurity but also for other impurity positions below the surface.

Since silver is a non-magnetic impurity the Kondo effect cannot be responsible for the observation. Furthermore, the overall phase shift induced by a Kondo impurity is stronger than observed for silver impurities. In the following, different phenomena which may produce such a signature will be discussed.

In general, the interference pattern at the surface of the Cu(100) surface can be related to properties of the host material (i.e. the band structure) and the scattering behavior of the impurity. Both quantities might be responsible for the kink in the interference pattern.

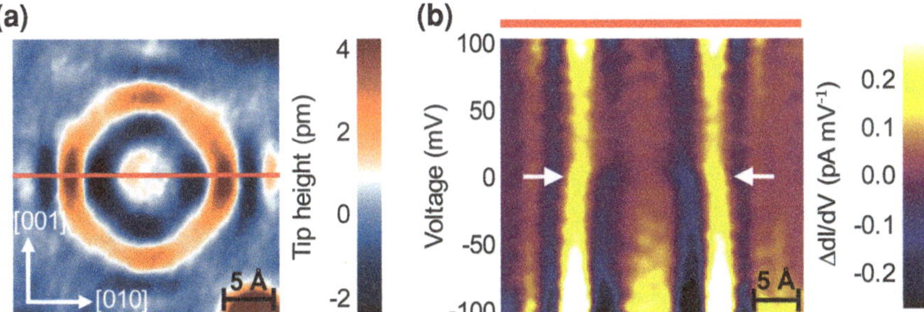

Fig. 5.4 Spectroscopic signature of a single silver atom around the Fermi energy. **a** Topography of a seventh monolayer Ag impurity (2.7 nm × 2.7 nm, −150 mV, 2 nA). **b** $\Delta dI/dV$ signal as function of applied bias voltage and one spatial coordinate, the [010] direction across the interference pattern (indicated by the *red line* in **a**). The tip height is adjusted to give a set point current of $I_{SP} = 2$nA at $V_{SP} = -300$ mV for the free surface. Around zero bias sample voltage a kink (highlighted by the *white arrows*) is observed in the interference pattern

From the impurity site the scattering behavior can be changed due to local vibrational modes. From STM experiments on adatom systems, like molecules on noble metal surfaces, it is well-known that vibrational modes can be excited by the tunneling current. The tunneling electrons can excite a vibrational mode, giving rise to an inelastic tunneling channel, which is forbidden when the energy of the electrons is below the quantized vibrational energy. This kind of inelastic tunneling contributions was first observed by Stipe et al. in 1998 [1] and leads to a symmetric step like feature in the differential conductance when the tip is placed above the molecule. An approach to understand how the surrounding charge oscillations, the Friedel-like oscillations, are affected by the localized vibrational mode was first provided by Fransson and Balatsky in 2007 [2] and more recently refined for inelastic spin excitations [3]. Experimentally the real space signature of Friedel-like oscillations was reported by Gawronski and collaborators in 2011 [4]. According to these works the scattering amplitude of the impurity should be enhanced for electron energies at the vibrational modes. An enhancement of the amplitude of the interference pattern around zero bias voltage is not found for embedded Ag impurities. One reason might be that not only one vibrational mode is involved but rather several modes with different vibrational energies are present.

Another reason could be the electron–phonon coupling in the copper crystal, meaning the coupling of the conduction electrons not to localized vibrational modes but to delocalized vibrational modes of the crystal lattice. This effect would not enter the scattering behavior of the impurity but the “band structure” of the underlying host material. T. Kotzott considered the influence of electron–phonon coupling on the spectral function of the itinerant electron system in the vicinity of a non-magnetic impurity in his Bachelor thesis [5]. The Einstein model as well as the Debye model is used to describe the properties of the phonon system. Both approximations result in a kink in the electron dispersion relation. In order to get the same strength as observed in the experiment the electron–phonon coupling parameter has to be chosen larger than values given in the literature. Moreover, the energy range of the kink is slightly smaller in the experiment than in the calculations.

5.4 Summary

In contrast to magnetic Co and Fe impurities, introduced in the previous chapter, the shift of the maxima and minima for non-magnetic Ag impurities is much weaker. For a broad energy range $V = \pm 300\text{mV}$ the development of the interference pattern can be understood by the concept of Friedel oscillations assuming that the scattering behavior of the impurity is nearly energy independent. However, high resolution spectroscopic data reveals a kink at the Fermi energy. Up to now the origin of this feature is unclear. One possible explanation might be the coupling of the conduction electrons either to localized vibrational modes present at the impurity site or to delocalized phonon modes arising from the crystal lattice. First theoretical calculations show that electron–phonon coupling in copper can produce a kink in the interference pattern. To rule out or to verify one model more experimental data as well as theoretical understanding of the different models and phenomena is required.

Using some information of the next chapter may give further insights. The interaction of two neighboring Fe atoms can lead to a suppression of the Kondo effect (see Sect. 6.2). In this particular case, the scattering behavior of the two Fe atoms becomes nearly energy independent (see Fig. 6.10). For the Fe dimer configuration no kink is found in the energy range of $V = \pm 30$ mV. If the kink in case of silver atoms is related to the electron–phonon coupling of the copper crystal it should also be observed in the interference pattern of the Fe dimer. This result shows that the electron–phonon coupling in the copper crystal cannot be responsible for the feature. More likely, the kink is related to localized vibrational modes present at the impurity or to an enhancement of the electron–phonon coupling due to specific impurity details (e.g. the mass).

During the measurements, it turned out that investigating Ag impurities seems to be a challenging task (compared to Fe or Co atoms). One reason might be that the valence properties of Ag are very similar to Cu ($3d^{10}4s^1$ and $4d^{10}5s^1$, respectively). Measuring the electrical resistivity is for a long time a standard way to determine the scattering properties of point defects. It was found that the absolute value of the resistivity does not only depend on the concentration of impurities but also on the impurity element. The results are summarized by Norbury's rule saying that the electrical resistivity increases and the scattering amplitude is proportional to the square of the valence difference of the partners [6]. Thus, germanium (Ge, $3d^{10}4s^2 4p^2$) or antimony (Sb, $4d^{10}5s^2 5p^3$) maybe good candidates to further investigate non-magnetic impurities in copper. In addition, measuring the spectroscopic signatures of a second non-magnetic element may give deeper insights into the underlying phenomena. If the feature is related to localized vibrational modes, the energy interval should change due to the different masses of Ag, Ge and Sb.

Although the silver samples were originally thought as reference system for Fe and Co impurities, the results go beyond the concept of simple Friedel oscillations. Moreover, point defects in metals are a very old topic in solid state physics [7, 8], which has regained strong attention for nanoscale systems [9–13]. While at room temperature electron–electron and electron–phonon scattering are dominating, the scattering of bulk electrons at single impurities becomes crucial at low temperature. Experimentally these systems have been investigated mainly by macroscopic transport (e.g. residual resistivity) or thermodynamic measurements (specific heat, magnetic susceptibility and thermal expansion) [6]. As a result of their limited spatial resolution these measurements always probe a very large ensemble of impurities. Real space techniques like high-resolution transmission electron microscopy (HRTEM) give access to structural changes due to the impurity. While the study of the electronic properties of a single point defect in metallic systems like an impurity in a host metal has been done to a large extent theoretically [14–16], experimental studies are rare due to the lack of techniques, which have the capability to investigate well-defined sample systems with high spectral and spatial resolution.

Basic theoretical concepts for the description of the change of the LDOS around an impurity (i.e. screening) are the well-known Thomas-Fermi and Friedel screening, which can be found in any modern solid state text book. While these theories treat the surrounding electron system as isotropic, it is well-known, for example from the giant magneto resistance (GMR) effect and the tunneling magneto resistance (TMR), that band structure effects play an important role for the understanding of the underlying physics in many systems.

A reason why single point defects have been rarely considered in the past is that they were assumed to have only a minor impact. While this might be justified for single impurities in a bulk system (neglecting many-body physics like the Kondo effect), it turns out that in nano devices and in mesoscopic systems devices like GMR and TMR structures small defect concentrations can dominate the overall behavior. The properties of interfaces are crucial to the functionality and quality of many electronic systems and devices [9–13]. The Cu(100) surface might be an excellent model system to investigate the impact of embedded non-magnetic atoms on the electronic structure of the surface from an experimental point of view.

References

1. Stipe BC, Rezaei MA, Ho W (1998) Science 280:1732
2. Fransson J, Balatsky AV (2007) Phys Rev B 75:195337
3. Fransson J, Balatsky AV (2012) Phys Rev B 85:161401
4. Gawronski H, Fransson J, Morgenstern K (2011) Nano Lett 11:2720

5. Kotzott T (2012) Signatur von Elektron-Phonon-Kopplung in STM-Spektren von nicht-magnetischen Atomen in Cu(100). Bachelor thesis, Georg-August-Universität zu Göttingen
6. Gottstein G (2004) Physical foundations of materials science. Springer, Berlin
7. Friedel J (1958) Il Nuovo Cimento 7:287
8. Ashcroft NW (1976) Solid state physics. Cengage learning services
9. Opitz J, Zahn P, Binder J, Mertig I (2001) Phys Rev B 63:094418
10. Wolf SA, Awschalom DD, Buhrman RA, Daughton JM, von Molnár S, Roukes ML, Chtchelkanova AY, Treger DM (2001) Science 294:1488
11. Mather PG, Read JC, Buhrman RA (2006) Phys Rev B 73:205412
12. Miller CW, Li Z-P, Akerman J, Schuller IK (2007) Appl Phys Lett 90:043513
13. Ke Y, Xia K, Guo H (2010) Phys Rev Lett 105:236801
14. Mertig I, Mrosan E, Schöpke R (1982) J Phys F: Met Phys 12:1689
15. Mertig I, Mrosan E, Zeller R, Dederichs PH (1983) Phys Stat Sol (b) 117:619
16. Braspenning PJ, Zeller R, Lodder A, Dederichs PH (1984) Phys Rev B 29:703

Chapter 6
Two-Impurity Kondo Physics

In small nanostructures containing only a few magnetic atoms as well as in periodically ordered systems the effective interaction between magnetic impurities leads to various magnetic ground states. In a metallic system, the magnetic spin orientation can be affected by direct exchange interaction due to an overlap of the localized impurity orbitals or by indirect exchange interaction mediated by the surrounding conduction electrons. Depending on the sign and nature of the interaction the system can exhibit ferromagnetic, antiferromagnetic or even more complex magnetic behavior such as canted spin structures. The interplay of these magnetic interactions with Kondo physics seems to be responsible for many phenomena, like unconventional superconductivity or quantum critical behavior, observed in heavy–fermion metals [1].

After the discussion on single iron and cobalt atoms in the previous chapter, this chapter will focus on the interaction between two neighboring iron impurities. These dimers show a considerably different spectral signature as compared to single impurities taken on the same sample. The Kondo resonance width is strongly altered or even suppressed due to the presence of the second iron atom. It turns out that the physical properties of the iron dimers strongly depends on their atomic configuration. At first, in Sect. 6.1 the properties of the two-impurity Kondo problem will be introduced. Then, in Sect. 6.2 two distinct dimer configurations are discussed in detail. The first configuration shows a broadening of the Kondo resonance for the dimer compared to a single impurity Fe atom. For the second configuration no spectroscopic signature of the Kondo effect is found in the experiment. Afterwards, an overview of different dimer configuration and their properties is presented. In Sect. 6.3 the experimental results are compared to recent simulations of sub-surface Fe dimer configurations showing fairly good agreement with the observed strength and directionality of the exchange interaction. Finally, in Sect. 6.4 the results are summarized followed by a short outlook for future experiments. Some parts of this chapter, in particular the calculation of the exchange interaction for distinct dimer configurations, originate from

H. Prüser, *Scanning Tunneling Spectroscopy of Magnetic Bulk Impurities*,
Springer Theses, DOI: 10.1007/978-3-319-06385-0_6

collaboration with Mohammed Bouhassoune and Samir Lounis from the Institute for Advanced Simulation at the Forschungszentrum Jülich.

Up to now only a few experiments discussed magnetic interactions between two atoms. This might be due to the lack of experimental techniques which have the capability to investigate well-defined sample systems with high spectral and spatial resolution. The interaction of two neighboring Co atoms on a Pt(111) surface was studied by measuring the magnetic hysteresis as function of their interatomic distance [2, 3]. More recently, the same method was used to measure the magnetic properties of small nanostructures built up of Fe atoms on a Cu(111) surface [4, 5]. This kind of magnetometry on the atomic scale might be the most direct way to investigate magnetic interaction, but the Kondo effect was not addressed in these works. First experiments treating the Kondo effect of coupled systems were done in the group of M. Crommie [6–8]. By using atomic manipulation nanostructures up to three atoms are built on an Au(111) surface. For Co dimers an abrupt disappearance of the Kondo effect can be seen for dimer distances less than 6 Å [6]. In case of Cr trimers it depends on the atomic configuration whether the Kondo effect is found or not [7]. Mapping the Kondo effect for Ni dimers as function of distance yields a reduced Kondo resonance width for interatomic distances less than 4 Å [8]. In all these experiments a change in the Kondo signature is only observed for very small interatomic distances, showing that the interaction energies for the (111) noble metal surfaces are rather small which is confirmed by recent calculations [9]. Connecting two Co atoms on Cu(111) by a Cu chain may enhance the coupling strength between the magnetic atoms resulting in an oscillation of the Kondo resonance width as function of the chain length [10]. So far, only a few experimental investigations considering the (100) orientation of noble metal surfaces exist. This rests upon the higher energy barriers of neighboring sites which make atomic manipulation on (100) noble metal surfaces very difficult. A well-defined and controlled way tailoring dimers on these surfaces is by self-assembly and manipulation of molecular complexes containing magnetic atoms [11, 12]. Using this method Co dimers on Cu(100) can be fabricated (see Fig. 6.1) showing a transition from ferromagnetic to antiferromagnetic coupling with increasing interatomic distance [11]. However, only for a dimer distance of 5.1 Å a significant broadening of the Kondo resonance width is found. For greater distances the same width as for a cobalt monomer is observed.

An alternative approach studying coupling of single magnetic atoms is by introducing a small insulating layer between the impurity and the metal surface. The decoupling opens the way to measure spin excitation on the atomic scale [13]. Investigating these kinds of systems the magnetism of a single atom [14], a dimer [15] or structures assembled of even more atoms [16] can be explored. However, decoupling reduces the Kondo temperature of the system making it hard to investigate these systems experimentally in the Kondo regime.

In recent years the coupling of two Kondo impurities was also investigated in quantum-dot experiments [17, 18]. Here, the advantage is that the parameters (e.g. coupling strength and Kondo temperature) can be tuned over a huge range. Due to

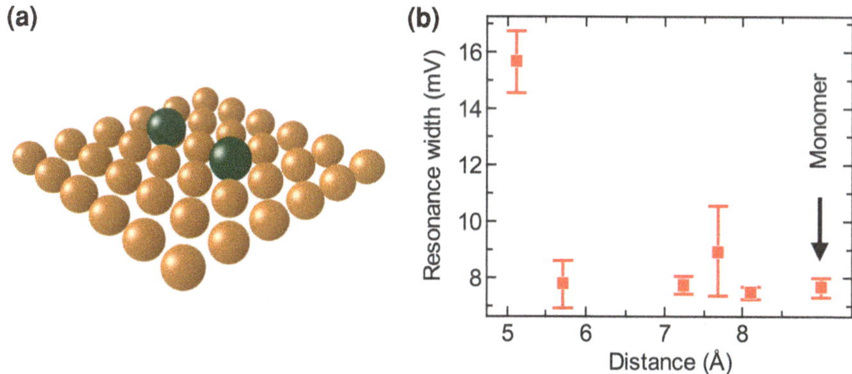

Fig. 6.1 Kondo resonance width of cobalt dimers on Cu(100). **a** Sketch of a magnetic dimer located on a crystalline surface. **b** Experimentally determined Kondo resonance width for different dimer distances. Reprinted with permission from [11]. Copyright 2007 by the American Physical Society. Only for a dimer distance of 5.1 Å (shown in **a**) a significant broadening of the Kondo resonance width is found. For larger distances the same width as for a cobalt monomer is observed

the geometric constraints such experiments cannot simply be expanded to more Kondo impurities.

A new concept for investigating two-impurity Kondo physics are atomic point contacts [19]. After picking up a Co atom from the surface to the tip of a scanning tunneling microscope, the tip itself becomes a Kondo system. Placing the tip above a surface magnetic atom allows studying the interaction. Although the coupling between the two atoms can be tuned continuously by varying the tip-sample separation, the results should be discussed carefully. Different interaction may occur driving the system into a point contact and also relaxation processes may happen changing the atomic arrangement during the measurement.

6.1 The Two-Impurity Kondo Model

The two-impurity Kondo model (TIKM) is the simplest model system containing local and non-local correlation phenomena resulting in a complex magnetic behavior discussed theoretically for a long time [20–23]. Local correlations arise from the coupling of each impurity to the surrounding conduction electrons. Since both impurities are coupled to the same electron bath non-local correlations naturally occur, leading to an interaction between both impurities. The localized spin exchange interaction of the impurities leads to a screening of the localized spins, the Kondo effect (discussed in a previous chapter). Despite this non-magnetic ground state, the interaction between the magnetic atoms may introduce either ferromagnetic or antiferromagnetic ordering. The introduction to the two-impurity Kondo model given in the following is mainly based on the article of Cho and McKenzie [24].

6.1.1 Fundamentals

The TIKM can be described by two localized spins interacting with the surrounding itinerant conduction electrons. The total system may be regarded as three subsystems, consisting of the two localized spins $\vec{S}_A$ and $\vec{S}_B$ which are separated by the distance R and surrounded by the conduction band electrons (see Fig. 6.2a).

The Hamiltonian for the two-impurity Kondo model is given by

$$H = \sum_{k,\sigma} \varepsilon_k c^{\dagger}_{k,\sigma} c_{k,\sigma} - J \cdot \vec{S}_A \vec{s}_c(A) - J \cdot \vec{S}_B \vec{s}_c(B) \tag{6.1}$$

The first part describes the non-interacting conduction electrons. The spin exchange interaction J describes the coupling between the localized impurity spins $\vec{S}_A$ and $\vec{S}_B$ with the conduction electron spin $\vec{s}_C$.

In general, the three sub systems interact both directly and indirectly with each other. For instance, both impurities interact directly with the conduction electrons by the spin exchange interaction J. This interaction leads to a screening of the localized spins, the Kondo effect. Since both impurities are coupled to the same electron bath they also interact indirectly, even though a coupling of both impurity spins does not appear explicitly in the model [24]. The information of the localized spin is passed on to the other impurity by the conduction electrons. This indirect interaction is well-known as Ruderman–Kittel-Kasuya-Yosida (RKKY) interaction [25–27]. Using second order perturbation theory in J this interaction can be described by the following model

$$H_{RKKY} = -J_{ex}(R)\vec{S}_A \cdot \vec{S}_B \tag{6.2}$$

where $J_{ex}(R) = 4\pi E_F J^2 F(2k_F R)$ is the effective spin exchange interaction of both impurities that depends on the distance R. For a three dimensional isotropic system the variation is given by $F_{3D}(x) = [\sin(x) - x\cos(x)]/x^4$, in case of lower dimensional systems the oscillation period is the same but the decay law differs. The spin exchange interaction in case of a three dimensional free electron system[1] is depicted in Fig. 6.2b. The interaction oscillates between ferromagnetic coupling (FM, $J_{ex}(R) > 0$) and antiferromagnetic coupling (AFM, $J_{ex}(R) < 0$). Ferromagnetic coupling always occurs for small distances due to the fact that both impurities couples to the same conduction electrons in the same way. However, for very small distances direct exchange interaction of the impurities, due to an overlap of the orbitals, gets more and more important and the model described above has to be modified.

[1] The parameters are chosen to yield an electron density which is comparable to a copper crystal, meaning $E_F = 7\,\mathrm{eV}$.

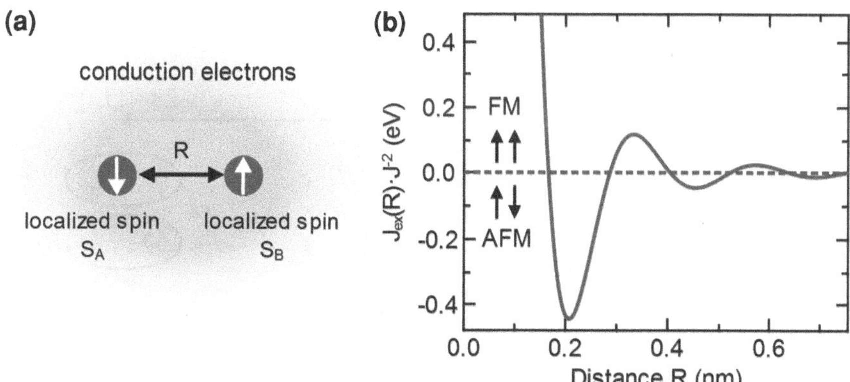

Fig. 6.2 The two-impurity Kondo model. **a** Scheme of the model, consisting of three sub systems: the two impurities separated by a distance R have a localized spin which interacts with the surrounding conduction electrons. Although there is no direct interaction between the impurities they interact indirectly by the RKKY interaction J_{ex} mediated by the conduction electrons. **b** Exchange interaction J_{ex} as function of distance R between the impurities for a three dimensional isotropic system

6.1.2 Competing Effects

In general, the ground state properties of the TIKM can be complicated. The thermodynamic properties of the model include a transition from a magnetically ordered state to a situation where the impurities behave like two independent Kondo impurities and the local moment is screened by the conduction electrons. The ground state properties can be derived by considering the two competing effects. On the one hand the Kondo effect tries to screen the impurity spin, resulting in a correlation of each impurity spin with the conduction electron spins but no spin-spin correlation of the impurities is found. The Kondo effect for a single impurity is defined by a characteristic energy scale Γ, which is determined by the Kondo temperature. On the other hand the RKKY interaction promotes a strong correlation between the impurity spins. By looking at the ratio $\Gamma/J_{ex}(R)$ of the Kondo energy scale with the spin exchange interaction $J_{ex}(R)$—describing the RKKY interaction—different scenarios are obtained, which are illustrated in Fig. 6.3.

If the Kondo energy scale is much higher than the RKKY interaction, the TIKM behaves, in the low temperature limit, like two uncoupled single impurities which are screened by the conduction electrons. The resulting Kondo resonance width of the two-impurity system Γ^* is the same as in the single-impurity Kondo model Γ. In case of non-negligible RKKY interaction, the crossover region, the exchange coupling induces either FM or AFM coupling between the impurities.

Up to now detailed investigations of the evolution of the spectral function for the crossover region are still missing. Recent results, calculating the spectral function of the localized orbital in the crossover region by perturbation theory,

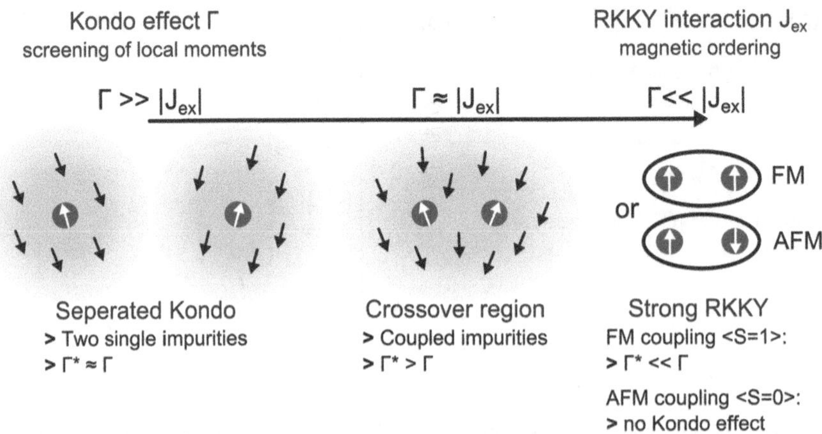

Fig. 6.3 Competition of the Kondo effect and RKKY interaction. If the Kondo effect at each impurity is much stronger than the RKKY interaction, the ground state can be described by two uncoupled and screened impurities. In the crossover region the RKKY interaction induces a coupling which leads to a broadening of the Kondo resonance. In case of strong RKKY interaction the impurity spins couple either ferromagnetically or antiferromagnetically resulting, respectively, in a two-step screening process with a strongly reduced resonance width or a vanishing Kondo effect

were obtained by Jabben et al. [23]. According to their work both FM and AFM coupling lead to an increase of the half-width at half maximum (HWHM) of the Kondo resonance in the spectral function of the localized orbital.

A schematic representation of the evolution of the spectral function for various AFM and FM interaction strength is shown in Fig. 6.4. For AFM coupling, see Fig. 6.4a, the broadening can be understood by the fact that the spectral weight of the Kondo resonance reduces for increasing interaction between the impurities. As a result, the half-width at half maximum (HWHM), the resonance width increases. For very large exchange coupling $J_{ex} < 0$ the two impurity spins align antiparallel $S = 0$ and no effective spin is seen by the conduction electrons resulting in a suppression of the Kondo effect and vanishing spectral signatures at the Fermi energy [20, 23].

For increasing FM coupling $J_{ex} > 0$ the spectral weight reduces, mimicking a broadening of the Kondo resonance, see Fig. 6.4b. In case of strong FM interaction the coupling leads to a parallel spin orientation acting as a rigid spin of total spin number $S = 1$. This effective higher spin value of the system gives rise to a two-step Kondo effect: By lowering the temperature of the system first one spin is screened and by further lowering the temperature the second spin is screened [20]. In the spectral function one narrow resonance appears [23]. The Kondo energy scale for the total impurity screening is reduced compared to the single impurity case $\Gamma^* < \Gamma$. This effect is comparable to multi-orbital Kondo physics increasing the spin of the system [28]. Normally the reduction is at least one order of magnitude which is usually lower than the experimental temperature. In this case the

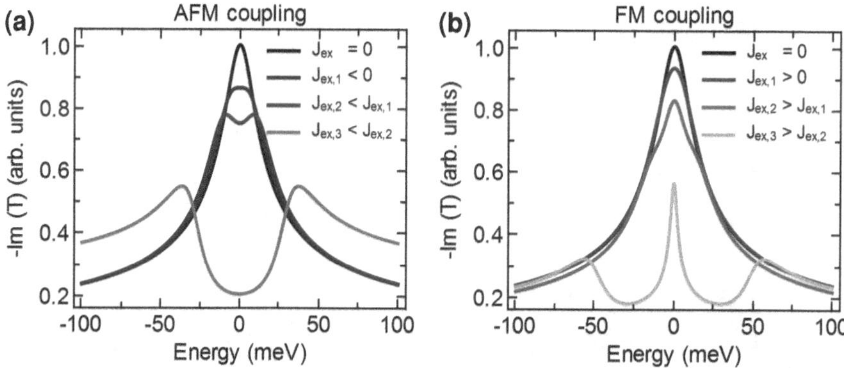

Fig. 6.4 Schematic representation of the spectral function of the localized orbital for the TIKM at zero temperature. **a** Evolution of the spectral function for increasing antiferromagnetic (AFM) coupling and **b** for increasing ferromagnetic (FM) coupling [23]. For small AFM and FM coupling a broadening of the resonance can be observed. However, for strong AFM exchange coupling a gap due to spin excitations is expected while for strong FM exchange coupling a narrow resonance survives at the Fermi energy

Kondo effect disappears since the system is not further in the low temperature limit, the Kondo regime. Experimentally, the scenario of a vanishing Kondo effect can be attributed to FM or AFM coupling only by temperature dependent measurements. For FM coupling, the Kondo effect should recur when the temperature is lowered [23] while for AFM coupling the Kondo effect stays absent.

6.2 Characterization of Fe Dimers

Iron dimer configurations can be identified by their topographic signature. When the Fe atoms are very close to each other the interference pattern starts to overlap. In general, the *bulk impurity* approach gives access to all possible dimer configurations in the crystal. A good starting point, however, is to study symmetric configurations in which the two impurities are located in the same monolayer below the surface. In these cases the atomic configuration of the dimer can be extracted by topographic measurements. Figure 6.5a shows standing-wave pattern of two neighboring Fe atoms. Comparing this with results found for single Fe impurities yields that they are located in the fifth monolayer. The center of each individual standing-wave pattern, marked by the red dots, reveals a displacement of 1.8 Å along the [010] direction and 5.4 Å along the [001] direction. These lengths correspond to an atomic configuration with interatomic distance $R = 0.57$ nm, which is depicted in Fig. 6.5b.

Although a topography measurement gives insights into the scattering behavior of the impurities, it has always the drawback of averaging the LDOS—the spectral function of the itinerant electrons—over a certain energy interval. To image this

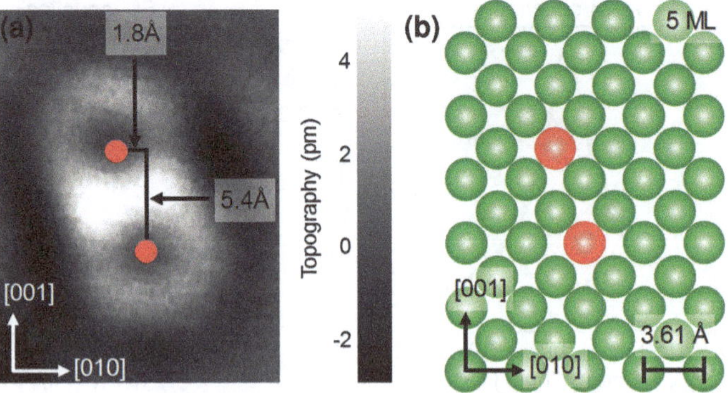

Fig. 6.5 Extracting the atomic configuration of two iron atoms close to each other. **a** STM constant current topography (1.8 nm × 2.3 nm, −31 mV, 0.2 nA) showing two neighboring fifth monolayer Fe impurities. The center of each standing-wave pattern, marked by the *red dots*, is displaced 1.8 Å and 5.4 Å, respectively, along the [010] and [001] direction. **b** Corresponding atomic configuration, the position of the Fe impurities is marked in *red*

quantity and to compare it to an isolated impurity the differential conductance is used in the following. Before going into a detailed analysis of the spectroscopic data acquired for different dimer configurations, the scattering behavior for two impurities will be discussed briefly.

One of the results of the previous chapter was that a single magnetic impurity embedded below a Cu(100) surface induces a complex $\Delta dI/dV(x, y, V)$ signature which depends on the lateral tip position (x, y) and the applied sample bias voltage V. The main idea for the description of bulk impurities is the separation of the electron propagation, determined by the host properties, and the scattering characteristics of the impurity, which is strongly affected by the Kondo effect and its characteristic energy scale Γ. A sketch of this concept is shown in Fig. 6.6a.

Due to the mirror symmetry the signature is the same for $\pm\alpha$, positive and negative angles of incidence with respect to the surface normal. The scattering behavior of two adjacent impurities, a dimer, can be described using the same concept, depicted in Fig. 6.6b, but is in general more complex. Since the propagation of the electrons and hence the spatial extension of the interference pattern on the surface of a single impurity is restricted, there exist three regions (I, II and III). In region I and III, where the electrons can only propagate to one impurity, to the left or to the right one, a signature $f'(x, y, \varepsilon)$ or $g'(x, y, \varepsilon)$ can be seen, respectively. Since the impurities are both Fe atoms, the scattering properties are equal. Note that the presence of a second impurity in general, leads to a changed scattering behavior of each impurity compared to the single impurity case $f \neq f' = g'$. If the Kondo effect is still present for the particular dimer configuration, the new scattering behavior is characterized by an altered energy scale Γ^*. The change of the resonance width provides information about the interaction of both atoms, discussed before in Sect. 6.1. In region II, electrons can propagate either to the left or to the right impurity and one obtains a superposition

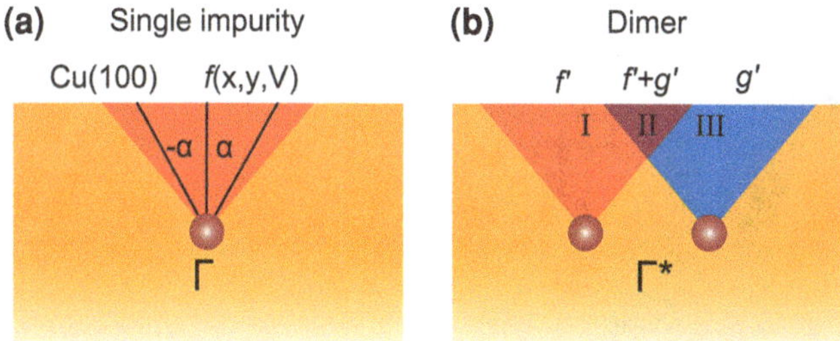

Fig. 6.6 Sketch of the superposition of two single impurity interference patterns. **a** Single embedded impurity, marked by the red circle induces a signature $f(x, y, \varepsilon)$ at the surface. Due to mirror symmetry the signature is the same for $\pm\alpha$. **b** Scattering behavior of a dimer. In general there exist three regions (I, II and III). In region I and III, where the electrons can only propagate to the left or to the right impurity, a signature $f'(x, y, \varepsilon)$ or $g'(x, y, \varepsilon)$ can be seen, respectively. Note that the presence of a second impurity leads to a changed scattering behavior (T-matrix) of each impurity compared to the single impurity case. In region II the interference pattern can be considered as a composition $f' + g'$ of one signature resulting from the left impurity f' and one signature resulting from the right g' impurity

$f' + g'$ of the interference patterns. Generally, the interpretation of single $\Delta dI/dV$ spectra in this area is quite complicated. In a previous chapter, see Sect. 4.7, it was shown that the spectroscopic signature of a single Kondo impurity, especially the line shape, strongly depends on the distance to the impurity. Since the distance to both impurities is not the same for most spatial positions in the superposition region (II), the contribution of each impurity, especially the line shape is different.

To extract microscopic parameters for magnetic dimers, the simplest approach is by single $\Delta dI/dV$ spectra obtained in region I and III or by spectroscopic measurements with a tip position direct at the center of the interference pattern. In the latter geometry the distance to both impurities is the same and the signal is twice as high as compared to equivalent positions in regions I and III. This special lateral tip position in region II provides spectroscopic data with better signal to noise ratio as for region I and III.

This first striking example, depicted in Fig. 6.7, shows that two adjacent Fe impurities show a considerably different spectral signature as compared to a single impurity. Since the measurements are performed with the same tip on the same sample, these measurements allow a direct comparison of a Fe dimer with an isolated single impurity located in the same monolayer without any artifacts or signatures which might be due to different electronic properties of the tip.

In the following, the above described concept of scattering for two adjacent impurities will be experimentally investigated. In Fig. 6.7 single $\Delta dI/dV$ spectra obtained for a Fe dimer ($R = 0.57$ nm) as well as for an isolated single impurity in the same layer (Fig. 6.7a and Fig. 6.7c), measured on the same sample with the same tip, are depicted. For a tip position direct at the center of the interference pattern of the dimer, see Fig. 6.7b, a narrow peak is found in the background subtracted differential conductance. Figure 6.7d clearly shows that for equivalent spatial positions on the interference pattern of the dimer and the monomer nearly

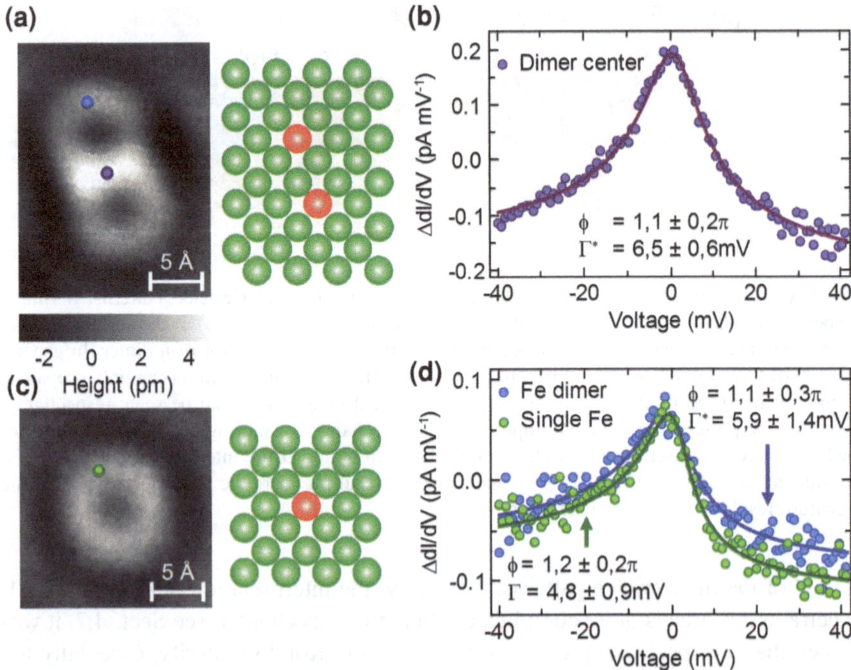

Fig. 6.7 Comparison of an iron dimer with an isolated single impurity located in the same layer. **a** STM constant current topographies (−31 mV, 0.2 nA) of the dimer with corresponding atomic configuration and **c**, the monomer as reference. Spectroscopic measurements were performed using a lock-in technique with a modulation of the sample voltage of $V_m = 1.7$ mV at a frequency of 2,190 Hz while the tip height is adjusted to give a set point current of $I_{SP} = 0.3$ nA at $V_{SP} = -40$ mV for the free surface. Single $\Delta dI/dV$ spectra for different lateral tip positions, marked in the topographies by circles. **b** Direct at the center of the interference pattern of the dimer a peak like feature is observed which shows a slightly higher Kondo resonance width as in the single impurity case. **d** More clearly this effect can be seen in spectra with the same lateral tip positions. The resonance width of the dimer is broadened, while the line shape as well as the amplitude is nearly the same for both spectra. The solid curves show fits to the spectra

the same line shape is observed. A closer look at the spectroscopic data reveals two main aspects. At first, the signature found in the center of the standing-wave pattern of the dimer is twice as high as the signature with a tip position at the border. This feature can be explained by the vicinity of the second Fe impurity which induces the same signature at the center as the first one and the overall signal can be considered as the sum. More significantly, the resonance width of the dimer is broadened, while the line shape as well as the amplitude is nearly the same for both spectra in Fig. 6.7d.

In analogy to the previous chapter about Kondo physics of single sub-surface atoms (see Chap. 4), the scattering behavior of a dimer, especially the phase shift, can be experimentally investigated by measuring the differential conductance as function of applied sample bias voltage and one spatial coordinate across the interference pattern. Figure 6.8 shows laterally resolved $\Delta dI/dV$ spectra for the above discussed dimer configuration with $R = 0.57$ nm. The section running across the center of the upper standing-wave pattern and perpendicular to the diagonal (Fig. 6.8b) as well as the section along the diagonal of the whole standing-wave pattern reveals an energy-dependent phase shift and enhanced amplitude around zero bias voltage. Furthermore, the enhanced amplitude at the center of the dimer interference pattern can be clearly seen. The laterally resolved $\Delta dI/dV$ signal of a single Fe impurity in the same layer along the same crystallographic direction, see Fig. 6.8d, e, reveals nearly the same voltage dependent evolution of the interference pattern.

One question naturally arising is how to extract microscopic information of the dimer in general. Up to now *there exists no fit formula or model for a magnetic dimer* in an STM experiment considering the electron propagation in the material and all different scenarios which may occur in the TIKM. Since the Kondo effect is still present in the above discussed dimer configuration and the single $\Delta dI/dV$ spectra as well as the $\Delta dI/dV$ sections look quite similar as for a single impurity, the simplest approach is to use the same fit formula introduced before for the single impurity case (see Sect. 4.5). Theoretically this model might be an appropriate approach for the crossover regime in the two-impurity Kondo model, dealing with intermediate exchange interaction J_{ex}. Here the spectral function of the localized orbital shows similar to the single impurity case a resonance at the Fermi energy which can be described by the phenomenological form found by Frota, however with a slightly increased resonance width, see Sect. 6.1.2. Fitting the spectroscopic signal of the dimer using Eq. (4.8) a Kondo resonance width of $\Gamma^* = 6.5 \pm 0.6$ meV and $\Gamma^* = 5.9 \pm 1.4$ meV is extracted with a tip position at the center and at the border, respectively. These values are slightly but significantly higher than the resonance width $\Gamma = 4.8 \pm 0.9$ meV which is observed for a single Fe impurity.

The behavior strongly depends on the atomic configuration with respect to the crystal lattice. In Fig. 6.9 the experimental results for a second dimer configuration, also located in the fifth monolayer, are shown. According to the topographic measurement, see Fig. 6.9a, b, the two Fe atoms are displaced by $R = 0.51$ nm along the $[01\underline{1}]$ direction with only one copper atom in between. While for the first dimer configuration, with $R = 0.57$ nm, the Kondo effect is still present but with an increased resonance width, for the second dimer configuration with $R = 0.51$ nm no Kondo fingerprints can be observed. Here single $\Delta dI/dV$ spectra measured at high symmetry points in the middle of the interference pattern (Fig. 6.9c, e) as well as spectra taken at the border (Fig. 6.9f) reveal a nearly linear voltage dependence and no signatures around zero bias at a sample temperature of 6 K.

Moreover, the laterally resolved $\Delta dI/dV$ signal, depicted in Fig. 6.10 shows no constriction of the interference pattern for positive bias voltage. The scattering behavior can be described by a constant phase and amplitude, no energy-dependent phase shift and no enhanced amplitude around zero bias voltage are observed. According to the theory the Kondo effect can vanish due to strong AFM or strong FM coupling.

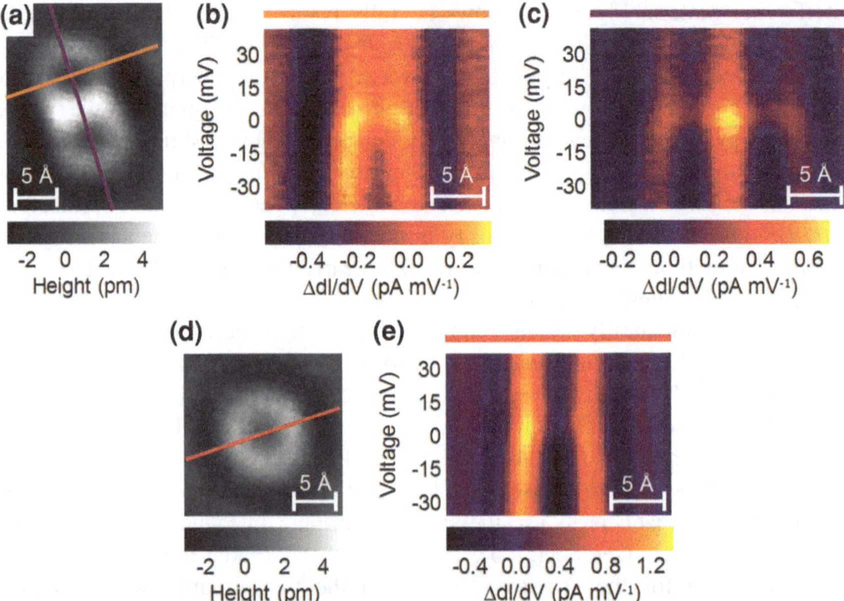

Fig. 6.8 Kondo signature of an iron dimer and one single impurity as continuous function of distance. **a** STM constant current topographies (−31 mV, 0.2 nA) of the dimer and **d** the monomer as reference. **b** $\Delta dI/dV$ signal as function of applied sample bias voltage and one spatial coordinate across the standing-wave pattern of the dimer (indicated by the orange line in **a**). **c** Section of the differential conductance along the diagonal of the dimer (indicated by the purple line in **a**). **e** $\Delta dI/dV$ section for an single iron impurity along the same direction as in **b** (indicated by the red line in **d**). All sections reveal an energy-dependent phase shift and increased amplitude of the interference pattern around zero bias voltage. Spectroscopic measurements were performed using a lock-in technique at a frequency of 2,190 Hz with a modulation of the sample voltage of $V_m = 1.7\,\text{mV}$ and $V_m = 2.1\,\text{mV}$ for the dimer and monomer, respectively. The tip height is adjusted to give a set point current for the free surface of $I_{SP} = 0.3\,\text{nA}$ and $I_{SP} = 0.5\,\text{nA}$ at $V_{SP} = -40\,\text{mV}$ and $V_{SP} = -36\,\text{mV}$ for the dimer and monomer, respectively

In order to get deeper insights into the interaction of two Kondo atoms with each other, Fig. 6.11 shows an overview of different iron dimer configurations with their corresponding spectroscopic signature. All atoms are located in the fifth monolayer and the interatomic distance between adjacent iron atoms ranges from $R = 0.51$ nm to $R = 1.02$ nm. For the most compact dimer configuration ($R = 0.51$ nm) observed so far in the experiment, the Kondo effect is suppressed at a sample temperature of 6 K. With increasing interatomic distance R features at the Fermi energy due to the Kondo effect are found. All the signatures can be fairly good described by the phenomenological fit formula introduced before. In Table 6.1 the resulting fit parameters, the resonance widths, the lineshape parameters as well as the resonance positions are summarized. As seen before for the single magnetic impurities in Sect. 4.7, the lineshape parameter increases for larger lateral tip positions. This behavior can be seen as a consistency check for the

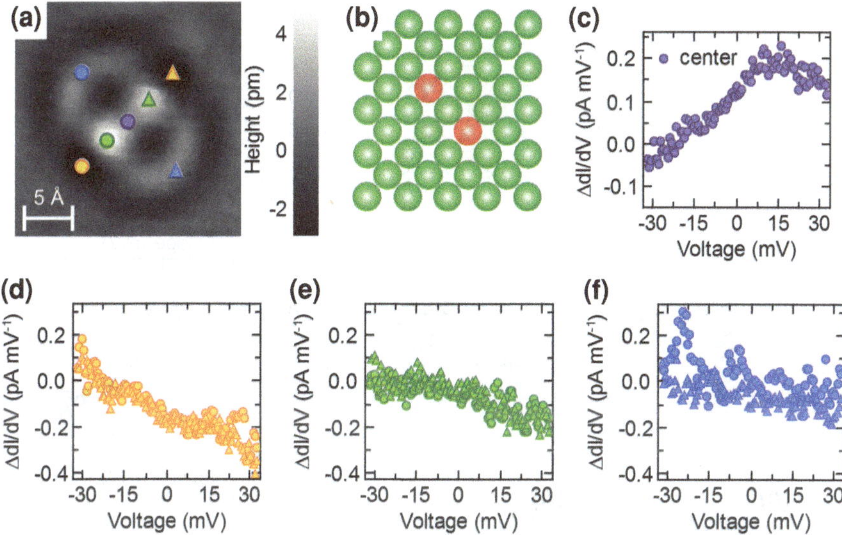

Fig. 6.9 Disappearance of the Kondo resonance for a compact dimer configuration. **a** STM constant current topography (2.3 nm × 2.3 nm, −31 mV, 0.2 nA) of the dimer and **b** corresponding atomic configuration. **c–f**, Single $\Delta dI/dV$ spectra for different lateral tip positions, marked in the topography. All spectra show no features around zero bias. The Kondo resonance observed for a single impurity is no longer present for this dimer configuration. Spectroscopic parameter: $I_{SP} = 0.25\,\text{nA}$, $V_{SP} = -31\,\text{mV}$ and $V_m = 1.4\,\text{mV}$ at a modulation frequency of 2,190 Hz

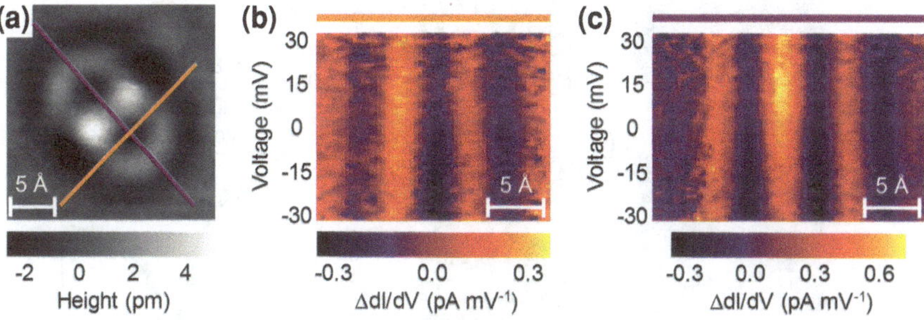

Fig. 6.10 Scattering behavior of a compact dimer configuration. **a** STM constant current topography (2.3 nm × 2.3 nm, −31 mV, 0.2 nA) of the dimer. **b** $\Delta dI/dV$ signal as function of applied sample bias voltage and one spatial coordinate across the standing-wave pattern of the dimer (indicated by the orange line in **a**). **c** Section of the differential conductance along the diagonal of the dimer (indicated by the purple line in **a**). Both sections reveal no energy-dependent phase shift and no increased amplitude of the interference pattern around zero bias voltage. Spectroscopic parameter: $I_{SP} = 0.25\,\text{nA}$, $V_{SP} = -31\,\text{mV}$ and $V_m = 1.4\,\text{mV}$ at a modulation frequency of 2,190 Hz

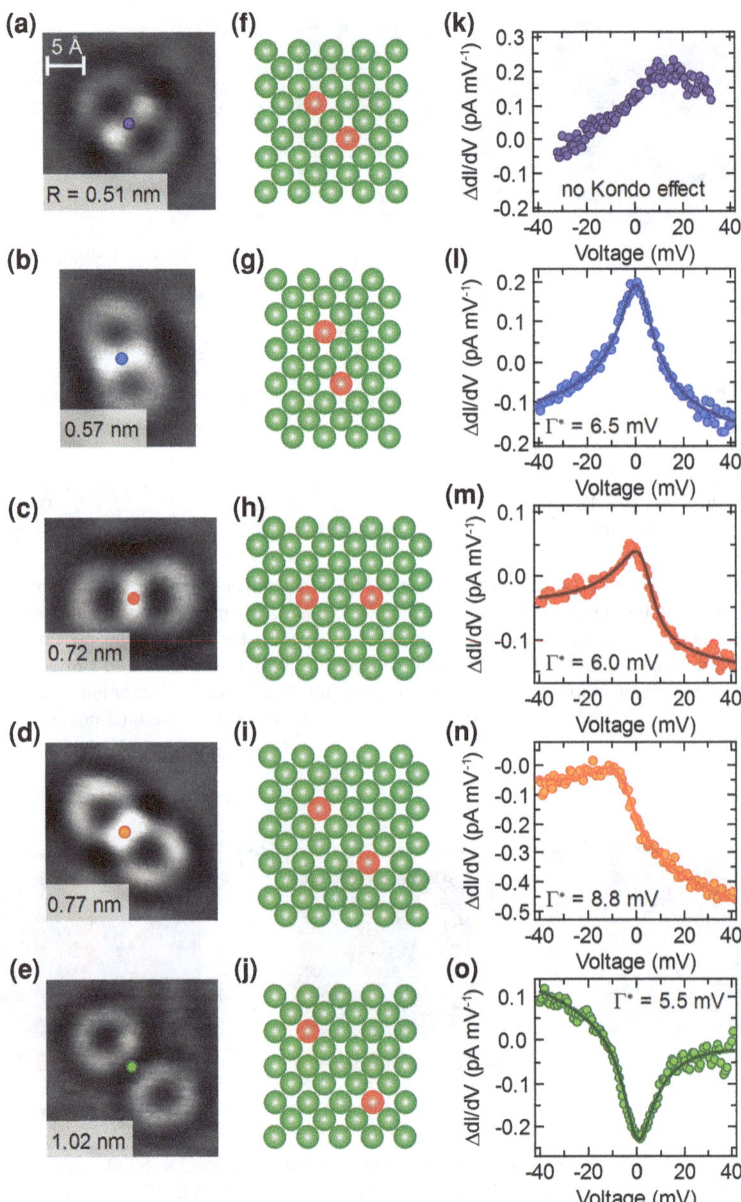

Fig. 6.11 Spectroscopic data for different Fe dimers. All dimers are situated in the fifth monolayer below the Cu(100) surface. **a–e** Topographies (−31 mV, 0.2 nA) of different dimer configurations with increasing distance. **f–j** The corresponding atomic configuration. **k–o**, Single $\Delta dI/dV$ spectra with a tip position at the center of the standing-wave pattern. While for a dimer distance of $R = 0.51$ nm no feature around zero bias are observed, all other configurations show Kondo fingerprints. The solid curves show fits to the spectra. Spectroscopic parameter: $I_{SP} = 0.25\,\text{nA}$ (**k**, **m**, **o**), $I_{SP} = 0.3\,\text{nA}$ (**l**, **n**), $V_{SP} = -40\,\text{mV}$ (**l–o**), $V_{SP} = -31\,\text{mV}$ (**k**) and $V_m = 1.4 - 1.7\,\text{mV}$ at a modulation frequency of 2,190 Hz

Table 6.1 Summary of the Fit parameter determined for different dimers

Configuration	Distance R (nm)	Γ^* (mV)	ϕ (π)	ε_k (mV)
3	0.51	No Kondo effect		
4	0.57	6.5 ± 0.6	1.1 ± 0.2	0.9 ± 0.9
5	0.72	6.0 ± 0.8	1.2 ± 0.2	3.2 ± 1.0
6	0.77	8.8 ± 1.2	1.3 ± 0.1	−3.9 ± 1.0
9	1.02	5.5 ± 0.6	1.9 ± 0.1	−0.1 ± 0.7
Single imp.	∞	4.6 ± 1.2		

The dimer configurations and their notation are depicted in the inset of Fig. 6.12. For comparison the resonance width for a single fifth monolayer Fe impurity ($\Gamma^* = \Gamma$), averaged over measurements with different tips and spatial positions, is also listed

scattering model of two neighboring atoms based on the separation of the scattering properties and the electron propagation.

The different configurations show a variation of the Kondo resonance width $\Gamma^*(R = 0.57\,\text{nm}) = 6.5 \pm 0.6\,\text{mV}$, $\Gamma^*(0.72\,\text{nm}) = 6.0 \pm 0.8\,\text{mV}$, $\Gamma^*(0.77\,\text{nm}) = 8.8 \pm 1.2\,\text{mV}$ and $\Gamma^*(1.02\,\text{nm}) = 5.5 \pm 0.6\,\text{mV}$. All values are increased compared to a single iron impurity which exhibits a resonance width of $\Gamma = 4.6 \pm 1.2\,\text{mV}$. [2]Also the experimental findings for single magnetic atoms that the Kondo resonance slightly increases for increasing lateral tip positions (see Sect. 4.8) cannot explain the high values. The observed behavior shows that there is an interaction between the iron atoms. In general, direct and indirect exchange mechanisms can be responsible for the interaction. If the localized orbitals overlap, the direct exchange interaction arises. This type of interaction always leads to an antiparallel alignment of the spins. Since the d-orbitals are quite localized, an overlap is expected only for a nearest neighbor configuration. The smallest interatomic distance observed so far in the experiment exhibits $R = 0.51$ nm. This is neither the nearest neighbor nor the next nearest neighbor configuration. Moreover, at least one copper atom is still in between the two iron atoms for the dimer configurations making the direct exchange coupling very unlikely to occur for the considered distances. More likely, the coupling results from the indirect RKKY interaction which is mediated by the conduction electrons.

Remarkably, the found resonance width is not a monotone function of the interatomic distance R. While for $R = 0.57$ nm and $R = 0.72$ nm the Kondo resonance is slightly increased compared to the single Fe impurity, the configuration $R = 0.77$ nm shows the strongest broadening. For a larger separation $R = 1.02$ nm the resonance width decreases and nearly the value for the single impurity is recovered. This clearly demonstrates that the interaction between two iron atoms not only depends on their interatomic distance but also depends on their

[2] The value for the single isolated fifth monolayer Fe impurity is obtained by averaging different measurements with different tips and lateral tip-distances.

atomic configuration with respect to the crystal lattice. This behavior can be understood by considering the anisotropy of the band structure of copper and the related electron focusing effect. This effect leads to charge oscillations which exhibit strongly enhanced intensities in certain directions and much weaker intensities in other directions. Therefore, the RKKY interaction, which is mediated by the spin charge density oscillation, should be also anisotropic.

6.3 Comparison with Theoretical Results

In order to check whether the experimental observations are related to a magnetic exchange interaction ab initio calculations have been developed by Mohammed Bouhassoune and Samir Lounis from the Institute for Advanced Simulation at the Forschungszentrum Jülich. The KKR (Korringa–Kohn–Rostoker) method within the local-density approximation (LDA) of the density functional theory (DFT) is used. As in the experiment, the iron impurities are located in the fifth monolayer below the Cu(100) surface. The magnetic exchange interaction has been extracted by mapping the ab initio calculations to the RKKY model, described in Eq. (6.2). Two magnetic configurations are considered: a ferromagnetic solution with energy $E(\uparrow\uparrow)$ where both magnetic moments are aligned parallel and an antiferromagnetic solution with $E(\uparrow\downarrow)$, where the magnetic moments are aligned antiparallel to each other. The exchange interaction J_{ex} is defined by the difference of both energies $J_{ex} = E(\uparrow\downarrow) - E(\uparrow\uparrow)$. In the following the exchange interaction J_{ex} is defined[3] by fixing the magnetic moment of each impurity to $\vec{S}_{A,B} = \pm 1 \cdot \vec{e}_r$. Figure 6.12 shows the calculated exchange interactions for the twelve nearest interatomic distances, the notation of the different configurations is depicted in the inset.

The resulting exchange interaction can be quantitatively understood by the RKKY model, introduced in Sect. 6.1.1. For the nearest neighbor configuration (#1) a strong ferromagnetic (FM) interaction with $J_{ex} = 73.1\text{meV}$ is predicted. The next nearest neighbor configuration (#2) exhibits an antiferromagnetic (AFM) alignment of the magnetic moments. The strength is strongly reduced compared to the first configuration (#1). For increasing distances the exchange interaction shows an oscillatory behavior, alternating between FM and AFM coupling. A closer look at the calculation of the exchange interaction reveals directions which show no oscillatory behavior. The most prominent candidate is the $[01\underline{1}]$ direction (green dots). Here the nearest neighbor configuration (#1) exhibits a FM coupling while all other dimer configurations (#3, #6 and #9) show an AFM coupling. This feature can be explained by an aliasing between the

[3] The exchange interaction J_{ex} is defined by the applied RKKY-Hamiltonian. For a total magnetic moment of $|S_{A,B}| = 2.7\mu_B$, which results from the ab initio calculation, the value of J_{ex} must be divided by a factor of 7.3. The direction of the magnetic moment is defined by $\vec{e}_r$.

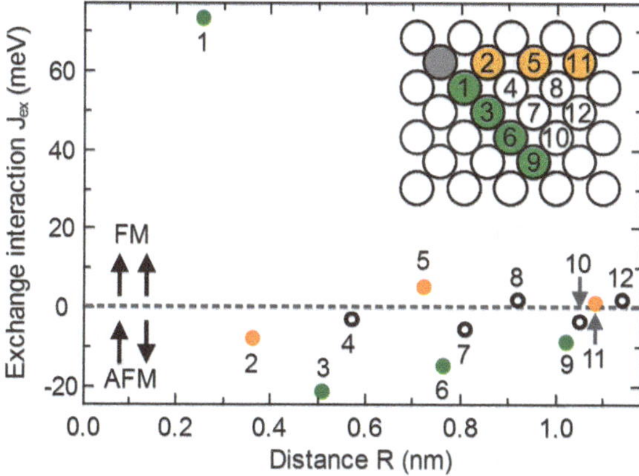

Fig. 6.12 Calculated exchange interaction as function of distance between iron atoms. While the impurity position below the surface is fixed to the fifth monolayer, the exchange interaction is calculated for various interatomic distances. The dimer notation, starting at #1 for the nearest neighbor configuration, #2 for the next nearest neighbor configuration and so on, is depicted in the inset. The [01$\underline{1}$] direction and the [010] direction are highlighted in *green* and *orange*, respectively

periodicity of the lattice and the wave length of the oscillations. For the [01$\underline{1}$] direction the lattice constant is 2.55 Å and the value of the Friedel wave length is $\lambda_{FW} = 2.47$ Å, according to the band structure calculation, presented in Sect. 3.1.3 . Not only the oscillation period is different for crystallographic directions but also the strength of the exchange coupling J_{ex}. Compared to the [010] direction (#2, #5 and #11, orange dots), the [01$\underline{1}$] direction (#1, #3, #6 and #9, green dots) offers a stronger exchange coupling. This shows that the exchange interaction is not isotropic but directional and strongly related to the band structure of the host material.

The first-principles results support the assumption that the different Kondo resonance width for the dimer configurations observed in the experiment are related to a magnetic interaction. Moreover, the calculation can reproduce the main features of the experimental findings. So far, the configurations (#3, #4, #5, #6 and #9) are experimentally investigated, see Sect. 6.2. In Fig. 6.13 the obtained Kondo resonance width Γ^* are directly compared with values for the exchange interaction predicted by the above described first-principles calculations. For the configuration $R = 0.51$ nm (#3) no Kondo signature around zero bias is found in the background subtracted differential conductance spectra. According to the simulation this configuration shows the strongest antiferromagnetic exchange coupling with $J_{ex} = -21.8\,\text{meV}$, all other AFM configurations exhibit smaller values. The coupling between the localized impurity spins seems to be much stronger than the coupling to the surrounding conduction electrons and the Kondo effect is suppressed. Going to the next configuration with slightly increased distance $R = 0.57$ nm (#4) the exchange coupling $J_{ex} = -3.4\,\text{meV}$ decreases and the Kondo effect recurs in the experiment. However, as discussed in Sect. 6.2, the observed resonance width $\Gamma^* = 6.5 \pm 0.6\,\text{mV}$ is considerably increased, compared

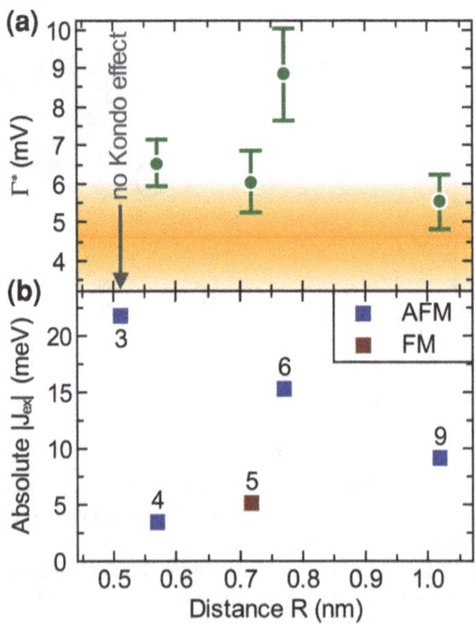

Fig. 6.13 Comparison of the experimental results with first-principles calculations. **a** Observed Kondo resonance width Γ^* as function of the interatomic distance R. As reference the averaged value for a single isolated fifth monolayer Fe impurity $\Gamma = 4.6 \pm 1.2\,\mathrm{mV}$ is highlighted by the orange line. **b** Absolute calculated values for the exchange interaction between two iron atoms below the Cu(100) surface. Ferromagnetic configurations marked in *red*, antiferromagnetic configurations in *blue*

to the single iron impurity Kondo resonance width $\Gamma = 4.5 \pm 1.0\,\mathrm{mV}$. According to the calculation, the second strongest AFM coupling with $J_{ex} = -15.2\,\mathrm{meV}$ is expected for configuration #6. Experimentally this configuration ($R = 0.77$ nm) exhibits the strongest broadening and an effective Kondo resonance width of $\Gamma^* = 8.8 \pm 1.2\,\mathrm{mV}$. The value is nearly doubled compared to the single impurity scenario. For the largest separation $R = 1.02$ nm (#9) only a small broadening is found $\Gamma^* = 5.5 \pm 0.6\,\mathrm{mV}$. Here the simulation predicts an AFM exchange interaction of $J_{ex} = -9.1\,\mathrm{meV}$ which is higher compared to the configuration #4. Therefore, the Kondo resonance width of configuration #9 should be larger as for #4 which is not the case according to the spectroscopic data. Here configuration #4 shows a stronger resonance broadening than configuration #9.

Following the simulation a ferromagnetic coupling is expected for configuration #5. An exchange interaction of $J_{ex} = 5.1\,\mathrm{meV}$ is predicted. According to the TIKM a FM exchange interaction can lead to a broadening or a reduction of the Kondo resonance width. Experimentally, an increased value $\Gamma^* = 6.0 \pm 0.8\,\mathrm{meV}$ is found. One possible explanation is that the FM exchange coupling is too small to produce a spin of $S = 1$, this is referred to as the crossover region (see Sect. 6.1.2). Another possibility is that the ferromagnetic interaction between the impurities is strong enough to produce a spin of $S = 1$ (strong RKKY regime) but the experimental temperature of 6 K is too high and the system is not in its ground state. As a consequence, the impurity spins are not fully screened which might lead also to an increased resonance width.

Comparing the experimental findings with the first-principles calculations demonstrates that the properties of the neighboring iron atoms can be related to the RKKY interaction. The calculation is able to reproduce the main features found in the experiment: The strength as well as the strong directionality of the magnetic interaction.

6.4 Conclusion

In conclusion, the results show that the scanning tunneling microscope cannot only be used to study single-impurity Kondo physics but also to investigate two-impurity Kondo physics of *bulk impurities* in real space. The experimental data is obtained on two adjacent iron atoms both located in the fifth monolayer below the Cu(100) surface. The interatomic distance between neighboring iron atoms ranges from $R = 0.51$ nm to $R = 1.02$ nm. As for a single impurity the interference pattern and the spectroscopic signature can be related to the scattering characteristics of the impurity and the electron propagation of the host material. However, for the case of two neighboring atoms the interference patterns overlap at the surface leading to different regions. In border regions and for positions with equal distances to both impurities the spectroscopic signature is described by the phenomenological fit formula introduced before for a single sub-surface Kondo impurity. Applying this fit formula shows that the Kondo resonance width Γ^* is strongly altered or even suppressed due to the presence of the second iron atom. For the smallest interatomic distance of $R = 0.51$ nm a vanishing Kondo effect can be observed. However, increasing the interatomic distance the Kondo effect recurs and the interaction between the impurities seems to enhance the Kondo resonance width. In addition, the behavior strongly depends not only on the interatomic distance but depends crucially on the atomic configuration of the dimer with respect to the host lattice. The strongest broadening is not observed for the smallest interatomic distance but for an intermediate separation, indicating an oscillatory behavior of the interaction energy.

The density functional calculations support the idea that the different dimer properties are related to the RKKY interaction mediated by the conduction electrons. The calculation reveals the same directional dependence as the measured data with a maximum in the AFM coupling strength for a distance of $R = 0.51$ nm along the $[01\underline{1}]$ direction. In both experiment and calculation this direction shows the strongest variation of J_{ex} with increasing interatomic distance, whereas it is weaker along the [010] direction. Since the RKKY interaction is mediated by the conduction electrons, the electron focusing effect could explain the observation. The results show that neighboring sub-surface magnetic atoms are a promising model system to study two-impurity Kondo physics. This chapter concentrates on the simplest case (at least for the analysis of the experiment) when the impurities are located in the same monolayer. In principle, the described method could also be used to study dimer configurations where the iron atoms are located in different layers.

The mapping of the scattering phase as a function of energy opens a new way to study the TIKM and to compare the experimental results with theory. Moreover, this new approach can also be used to investigate larger magnetic atom assemblies containing more than two impurities which are coupled to each other. Being able to measure the magnetic coupling for two impurity atoms is a prerequisite for the understanding and design of more complex structures. For example trimers or linear chains are promising model systems for a bottom up approach towards materials where the local moments are arranged in a periodic array, so-called Kondo lattices.

References

1. Gegenwart P, Si Q, Steglich F (2008) Nat Phys 4:186
2. Meier F, Zhou L, Wiebe J, Wiesendanger R (2008) Science 320:82
3. Zhou L, Wiebe J, Lounis S, Vedmedenko E, Meier F, Blügel S, Dederichs PH, Wiesendanger R (2010) Nat Phys 6:187
4. Khajetoorians AA, Wiebe J, Chilian B, Wiesendanger R (2011) Science 332:1062
5. Khajetoorians AA, Wiebe J, Chilian B, Lounis S, Blügel S, Wiesendanger R (2012) Nat Phys 8:497
6. Chen W, Jamneala T, Madhavan V, Crommie MF (1999) Phys Rev B 60:R8529
7. Jamneala T, Madhavan V, Crommie MF (2001) Phys Rev Lett 87:256804
8. Madhavan V, Jamneala T, Nagaoka K, Chen W, Li J-L, Louie SG, Crommie MF (2002) Phys Rev B 66:212411
9. Simon E, Újfalussy B, Lazarovits B, Szilva A, Szunyogh L, Stocks GM (2011) Phys Rev B 83:224416
10. Néel N, Berndt R, Kröger J, Wehling TO, Lichtenstein AI, Katsnelson MI (2011) Phys Rev Lett 107:106804
11. Wahl P, Simon P, Diekhöner L, Stepanyuk VS, Bruno P, Schneider MA, Kern K (2007) Phys Rev Lett 98:056601
12. Wegner D, Yamachika R, Zhang X, Wang Y, Baruah T, Pederson MR, Bartlett BM, Long JR, Crommie MF (2009) Phys Rev Lett 103:087205
13. Heinrich AJ, Gupta JA, Lutz CP, Eigler DM (2004) Science 306:466
14. Hirjibehedin CF, Lutz CP, Heinrich AJ (2006) Science 312:1021
15. Otte AF, Ternes M, Loth S, Lutz CP, Hirjibehedin CF, Heinrich AJ (2009) Phys Rev Lett 103:107203
16. Loth S, Baumann S, Lutz CP, Eigler DM, Heinrich AJ (2012) Science 335:196
17. Jeong H, Chang AM, Melloch MR (2001) Science 293:2221
18. Craig NJ, Taylor JM, Lester EA, Marcus CM, Hanson MP, Gossard AC (2004) Science 304:565
19. Bork J, Zhang Y-H, Diekhöner L, Borda L, Simon P, Kroha J, Wahl P, Kern K (2011) Nature Phys. 7:901
20. Jayaprakash C, Krishna-murthy HR, Wilkins JW (1981) Phys Rev Lett 47:737
21. Jones BA, Varma CM (1987) Phys Rev Lett 58:843
22. Jones BA, Varma CM, Wilkins JW (1988) Phys Rev Lett 61:125

23. Jabben T, Grewe N, Schmitt S (2012) Phys Rev B 85:045133
24. Cho SY, McKenzie RH (2006) Phys Rev A 73:012109
25. Kasuya T (1956) Prog Theor Phys 16:45
26. Ruderman MA, Kittel C (1954) Phys Rev 96:99
27. Yosida K (1957) Phys Rev 106:893
28. Nevidomskyy AH, Coleman P (2009) Phys Rev Lett 103:147205

Chapter 7
Outlook

Since the first STM studies in 1998 [1, 2] showing Kondo signatures many experiments on magnetic atoms and molecules *on metal surfaces* have been carried out. With its unprecedented control the STM technique allows for the first time to investigate Kondo fingerprints that occur on very small length and energy scales. Most of the studies focused on the Kondo signature, its dependence on the local environment [3–8] for example step edges and the interaction between two or even more Kondo impurities. The great advantage of adatom systems is their rather huge Kondo signature (up to 30 % of the signal [2]) and that atomic manipulation techniques can be used to fabricate artificially magnetic nanostructures. All of the adatom studies give further insights into the microscopic description of Kondo physics and regained interest on single magnetic impurities. Nevertheless, some of the key questions of Kondo systems the spatial extension of the Kondo effect and the interaction between magnetic atoms are still under debate. Up to now Kondo features are just found when the tip is placed directly above the impurity position. In addition, an interaction between impurities is only observed for very small interatomic distances (mostly for nearest neighbor configurations). The missing success of observing spatially extended Kondo signatures and interactions between impurities for surface magnetic atoms and molecules is up to now not fully understood and still under discussion. An ongoing debate concerns the influence of the surface state on the Kondo effect [3, 5, 9–11].

In this work an alternative approach is exploited—atoms below a metal surface. It turns out that bulk impurities can be investigated by STM on relatively long distances from the impurity opening a promising new way to Kondo physics in real space. Single magnetic Fe and Co impurities embedded below the Cu(100) surface show a complex spatially dependent signature. The analysis shows that magnetic sub-surface impurities offer a hitherto unobserved spatially extended Kondo signature which allows determining the phase shift. This quantity of the Kondo effect cannot be extracted from a single dI/dV curve (magnetic adatoms) or from PES/IPES measurements.

H. Prüser, *Scanning Tunneling Spectroscopy of Magnetic Bulk Impurities*, Springer Theses, DOI: 10.1007/978-3-319-06385-0_7

The main idea for the description of Kondo bulk impurities is the separation of the electron propagation, related to the band structure of the host material, and the scattering characteristics of the impurity, which is influenced by the Kondo effect. Within this interpretation, the STM is a noninvasive tool that exclusively probes the LDOS of the conduction band electrons—the itinerant electrons. This is in contrast to adatom systems. Here, theoretical models discuss the line shape of the Kondo signature in terms of interfering channels considering tunneling into the localized orbital and the conduction band states, respectively [12, 13]. Thus for adatom systems, the line shape is a feature of the measurement process and is associated with the relative strength of the hybridization between the tip, the sample and the impurity.

In order to get further insights into the properties of magnetic impurities a magnetic field could be used. Recently, a splitting of the Kondo resonance by an applied magnetic field was reported for cobalt and titanium atoms adsorbed on Cu_2N [14]. A magnetic field should also affect the scattering behavior of subsurface impurities. Available magnetic fields of a few Tesla are too low to observe a significant splitting or broadening for iron or cobalt atoms in copper, but manganese (Mn) may be a good candidate for such an experiment.

The new approach cannot only be used to study single-impurity Kondo physics but also to investigate two-impurity Kondo physics. In general, the two-impurity Kondo model reveals a rich physical scenario, including a crossover from local Kondo screening of the constituents to the formation of magnetic ordering, which can be either antiferromagnetic- or ferromagnetic-like. Depending on the interatomic distance, all different scenarios can be observed experimentally for iron dimers buried below a Cu(100) surface.

Furthermore, bulk impurities are well-known to show *universal behavior* to a certain extent. The results for one magnetic species—here iron and cobalt in copper—allows a prediction for other magnetic elements in the same host material.

According to the Kondo model (s-d model, see Sect. 4.1) the resonance width $\Gamma \propto \exp(-1/2J\rho)$, should depend on the strength of the exchange coupling J and the density of the conduction electron states $\rho\left(x_{imp}, E_F\right)$ at the Fermi energy and at the magnetic impurity position x_{imp} [15]. In contrast, the RKKY interaction $J_{ex} \propto J^2$ is proportional to the square of the exchange coupling J (see Sect. 6.1.1). The interplay between the two interactions was first emphasized by S. Doniach in 1977 [16]. The evolution of the Kondo resonance width Γ (blue curve) and the RKKY interaction J_{ex} (green curve) as function of the spin exchange parameter J is illustrated in Fig. 7.1.[1] This kind of plot, known as Doniach diagram, illustrates that for small values of J the RKKY interaction is stronger than the Kondo effect leading to either ferromagnetic or antiferromagnetic ordering. For high values of J the Kondo interaction dominates over the RKKY interaction resulting in screened impurity spins this is referred to as the Kondo regime. According to the two-impurity Kondo model the ground state properties of the system are

[1] For simplicity the density of the conduction electron states $\rho\left(x_{imp}, E_F\right) = 1$ as well as all other constants are set to one.

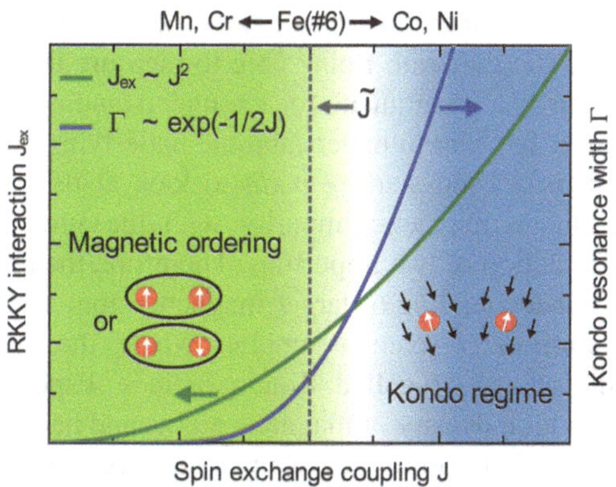

Fig. 7.1 Doniach diagram for the two-impurity Kondo model. The Kondo resonance width Γ (*blue curve*), indicating the Kondo energy scale, and the RKKY interaction J_{ex} (*green curve*) are shown as function of the spin exchange coupling J. When J is small the RKKY interaction $J_{ex} > \Gamma$ is stronger than the Kondo effect leading to a magnetic ordering while for high values of J the Kondo effect screens the local moments $J_{ex} < \Gamma$. A given dimer configuration can be characterized by an effective spin exchange coupling $\tilde{J}$ showing the same ratio of Γ/J_{ex}. For the same dimer configuration built up of manganese (Mn) or chromium (Cr) a reduction of the effective spin coupling is expected. In contrast, for cobalt (Co) and nickel (Ni) dimers, the effective spin coupling should be higher

determined by the ratio Γ/J_{ex} of the single-impurity Kondo resonance width Γ and the RKKY interaction J_{ex}. A given Fe dimer configuration, for example the configuration with $R = 0.77$ nm (#6, see Fig. 6.11) indicating a strong broadening of the Kondo resonance width, can be attributed to a scenario where the RKKY interaction and the Kondo energy scale are of the same order (see Sects. 6.2 and 6. 3). The configuration can be described by an effective spin coupling $\tilde{J}$ in the Doniach diagram (grey dashed line) showing the same ratio of Γ/J_{ex}. Macroscopic measurements yield an increasing Kondo resonance width and hence increasing spin exchange coupling J for manganese (Mn), chromium (Cr), iron (Fe), cobalt (Co) and nickel (Ni) in copper (Cu) [17]. This implies that for same dimer geometry (for example #6) for Mn and Cr the RKKY interaction should be more prominent while for Co and Ni impurities the Kondo effect should dominate. In the diagram the effective spin exchange for a distinct dimer configuration moves to the left for Mn and Cr dimers and to the right for Co and Ni, respectively. This scaling predicts for Cr and Mn dimers a much stronger RKKY interaction as for Fe dimers. With a base temperature low enough the RKKY interaction should be experimentally observable at longer distances for these systems.

Moreover, the results are not only relevant for nano structured systems or surface science but also for the investigation of macroscopic systems. It was shown that single magnetic impurities resemble bulk properties. This shows that the STM technique allows for the first time to study phenomena that occur on very small length and energy scales and to investigate how they can be related to the macroscopic properties of the material. Being able to measure the magnetic coupling for two impurity atoms is a prerequisite for the understanding and design of more complicated structures. In order to understand the underlying physical concepts of a complex system it is a good starting point to look at its building blocks: the interaction of one single magnetic impurity with the surrounding conduction electrons and the interaction of two impurities. Therefore, the results are a bottom-up approach for the understanding of larger magnetic atom assemblies containing more than two impurities or even assemblies where the magnetic atoms are arranged in periodic arrays, so-called Kondo lattices. Especially for the latter systems which have a high density of magnetic atoms and hence small interatomic distances the interaction between impurities plays a crucial role for the electronic properties of the system. At low temperature these systems develop unusual magnetic, superconducting or orbital ordering. How these properties emerge and how they are related to microscopic structure or defects is largely an open question. The achievement of high purity single crystalline samples as well as the developing field of STM pave the way to investigate the connection between collective macroscopic phenomena studied in the past and the structural (defects and vacancies) and electronical properties down to the atomic scale [18–22]. Here the capability of STM to investigate *bulk impurities* may lead to a broader understanding of these materials and their theoretical description.

Finally, the observation of a spatially extended Kondo signature may give an insight into one of the most controversial terms discussed in Kondo physics: the meaning and size of the "Kondo cloud". The "Kondo cloud" describes a theoretical concept, considering the length scale ξ_k of correlations in a Kondo system. Typically, this value is in the range of 100–1000 nm. Naturally the best way to visualize ξ_k is by measuring the spin–spin correlation function. Up to now there exists no experimental technique to investigate this property. Recent theoretical calculations show that the charge density oscillation $n(r)$ can be related to the correlation length scale [23]. The charge density can be in principle obtained in a STM experiment through integrating the $\Delta dI/dV$ spectra over a broad energy interval from the band onset to the Fermi energy. Although this might be possible, for high voltages the $\Delta dI/dV$ signal is no longer proportional to the energy resolved density of states. Here the electronic structure of the tip as well as the transmission coefficient cannot be further neglected. Nevertheless, a property which can be seen in the charge density should also show up in the energy resolved charge density—the LDOS. With the long–range LDOS signatures of buried impurities the term 'Kondo cloud' may be defined in a way that is observable and consistent with the experiments. However, the observed spatial extension of the Kondo signature in the experiment only a few nm is still too short. Going to materials with more pronounced electron focusing and increasing the experimental resolution might pave the way to measure the "Kondo cloud".

References

1. Li J, Schneider W-D, Berndt R, Delley B (1998) Phys Rev Lett 80:2893
2. Madhavan V, Chen W, Jamneala T, Crommie MF, Wingreen NS (1998) Science 280:567
3. Henzl J, Morgenstern K (2007) Phys Rev Lett 98:266601
4. Iancu V, Deshpande A, Hla S-W (2006) Phys Rev Lett 97:266603
5. Limot L, Berndt R (2004) Appl Surf Sci 237:572
6. Uchihashi T, Zhang J, Kröger J, Berndt R (2008) Phys Rev B 78:033402
7. Néel N, Kröger J, Limot L, Palotas K, Hofer WA, Berndt R (2007) Phys Rev Lett 98:016801
8. Néel N, Kröger J, Berndt R, Wehling TO, Lichtenstein AI, Katsnelson MI (2008) Phys Rev Lett 101:266803
9. Knorr N, Schneider MA, Diekhöner L, Wahl P, Kern K (2002) Phys Rev Lett 88:096804
10. Schneider M, Vitali L, Wahl P, Knorr N, Diekhöner L, Wittich G, Vogelgesang M, Kern K (2005) Appl Phys A Mater Sci Process 80:937
11. Li Q, Yamazaki S, Eguchi T, Kim H, Kahng S-J, Jia JF, Xue QK, Hasegawa Y (2009) Phys Rev B 80:115431
12. Schiller A, Hershfield S (2000) Phys Rev B 61:9036
13. Plihal M, Gadzuk JW (2001) Phys Rev B 63:085404
14. Otte AF, Ternes M, von Bergmann K, Loth S, Brune H, Lutz CP, Hirjibehedin CF, Heinrich AJ (2008) Nature Phys 4:847
15. Hewson AC (1993) The Kondo problem to heavy fermions. Cambridge University Press, Cambridge
16. Doniach S (1977) Physica B + C, 91:231
17. Daybell MD, Steyert WA (1968) Rev Mod Phys 40:380
18. Aynajian P, da Silva Neto EH, Parker CV, Huang Y, Pasupathy A, Mydosh J, Yazdani A (2010) Proceedings of the National Academy of Sciences 107:10383
19. Schmidt AR, Hamidian MH, Wahl P, Meier F, Balatsky AV, Garrett JD, Williams TJ, Luke GM, Davis JC (2010) Nature 465:570
20. Ernst S, Kirchner S, Krellner C, Geibel C, Zwicknagl G, Steglich F, Wirth S (2011) Nature 474:362
21. Wirth S, Ernst S, Cardoso-Gil R, Borrmann H, Seiro S, Krellner C, Geibel C, Kirchner S, Burkhardt U, Grin Y, Steglich F (2012) J Phys: Condens Matter 24:294203
22. Aynajian P, da Silva Neto EH, Gyenis A, Baumbach RE, Thompson JD, Fisk Z, Bauer ED, Yazdani A (2012) Nature 486:201
23. Affleck I, Borda L, Saleur H (2008) Phys Rev B 77:180404

References

1. [illegible] Schmiedmayer [illegible] (1998) Phys Rev Lett [illegible]
2. [illegible] (1998) Science [illegible]
3. [illegible] Phys Rev Lett [illegible]
4. [illegible]
5. [illegible]
6. [illegible] Phys Rev B [illegible]
7. [illegible] (2007) Phys Rev [illegible]
8. [illegible] Phys Rev [illegible]
9. [illegible]
10. [illegible] (2005) [illegible]
11. [illegible] Phys Rev [illegible]
12. [illegible]
13. [illegible] Phys Rev B [illegible]
14. [illegible] (2008) Nature Phys [illegible]
15. Hewson AC (1993) The Kondo problem to heavy fermions. Cambridge University Press, Cambridge
16. Doniach S (1977) Physica B [illegible]
17. [illegible] (1984) Rev Mod Phys [illegible]
18. [illegible] (2010) [illegible]
19. [illegible] (2010) Nature [illegible]
20. [illegible]
21. [illegible] (2012) J Phys Condens Matter 24 [illegible]
22. [illegible] (2012) Nature [illegible]
23. [illegible] Phys Rev [illegible]

Chapter 8
Summary

In this work the transition from single *bulk Kondo systems* towards *interacting Kondo atoms* is investigated using low temperature (6 K) scanning tunneling microscopy and spectroscopy. The investigation of single atoms buried below a Cu(100) surface requires well-defined samples, which contain a low concentration of impurities. For this purpose samples with silver (Ag), iron (Fe) and cobalt (Co) atoms are prepared through simultaneous deposition of copper and the impurity material from two electron beam evaporators under UHV conditions. Although this seems to be straightforward a lot of experimental challenges have been solved. In order to avoid segregation to the surface during growth each impurity material requires a different sample preparation scheme. During the last years the sample quality as well as the electronic setup for the high resolution spectroscopic data acquisition has been continuously improved.

This type of samples allows studying single-impurity Kondo physics as well as interacting Kondo systems in real space for impurity positions down to the seventh monolayer. Spatially dependent low temperature spectroscopic information is used to map the itinerant electrons—the conduction electrons at the surface in the vicinity of sub-surface impurities. Magnetic Fe and Co atoms induce a long range Kondo signature. Ag atoms have been studied as a non-magnetic reference system proving that the signatures are related to the localized magnetic moment.

Universal properties of the Kondo effect are demonstrated by comparing Co and Fe atoms. Both impurity species show similar behavior on completely different energy scales which can be understood by their different mean Kondo temperatures of $T_k = 37 \pm 12\,\mathrm{K}$ and $T_k = 655 \pm 155\,\mathrm{K}$ for Fe and Co atoms, respectively. A theoretical model treating the Kondo effect as a scattering resonance right at the Fermi energy is able to describe the measured signatures.

In comparison to STM experiments on magnetic molecules or atoms on noble metal surfaces the embedded impurities show a complex spatially dependent signature. This allows a mapping of the scattering phase caused by the Kondo resonance. By analyzing the scattering phase shift as function of energy it turned

H. Prüser, *Scanning Tunneling Spectroscopy of Magnetic Bulk Impurities*,
Springer Theses, DOI: 10.1007/978-3-319-06385-0_8

out that the often-used Lorentzian approximation of the Kondo resonance, resulting in the well-known Fano line shapes, does not give the best description. A phenomenological form found by Frota provides a better model of the Kondo resonance and hence the differential conductance. While the discrepancy of both models can hardly be observed for a single spectrum, the difference can be clearly seen by analyzing the phase shift. The experiments give a value of the order of $\pi/2$ while the Lorentzian approximation of the Kondo resonance always results in a phase shift of π. Recent many-body calculations also show that the phase shift is strongly overestimated by the Lorentzian approximation. Modeling the Kondo resonance by the phenomenological form found by Frota yields a smaller phase shift of only $\pi/2$, which is of the same order as observed in theory and experiment.

A detailed analysis of the differential conductance as function of lateral tip position and impurity distance below the surface is presented for both magnetic impurity atoms. The line shape of the Kondo signature shows an oscillatory behavior as function of distance, which has been predicted in the past but so far was not observed in the experiment. The oscillation period for the different directions are anisotropic, which is in a very good agreement with the band structure of the underlying copper crystal.

Apart from isolated single impurities, geometries where two Fe impurities are located in the same monolayer close to each other are investigated. The dimer configurations show a considerably different spectral signature as compared to single Fe impurities located in the same monolayer. The Kondo effect, in particular the resonance width is strongly altered or even suppressed due to the presence of the second Fe atom. It is shown that the behavior is related to the two-impurity Kondo model comprising two competing effects: On the one hand the Kondo effect, which tries to screen the impurity spin. On the other hand the RKKY interaction promoting a strong correlation between the magnetic moment of the two impurities. The investigation of different dimer geometries with interatomic distances ranging from $R = 0.51$ nm to $R = 1.02$ nm show that the behavior strongly depends on the atomic configuration of the iron dimer and not simply on the geometric distance. For the smallest interatomic distance a vanishing Kondo effect can be observed. Increasing the interatomic distance the Kondo effect recurs and the interaction between the impurities enhances the Kondo resonance width. The strongest broadening does not occur for the smallest interatomic distance but for an intermediate separation, indicating an oscillatory behavior of the RKKY interaction.

Density functional theory calculations support the idea that the dimer properties are related to the RKKY interaction mediated by the conduction electrons. The *ab initio* calculation reveals the same directional dependence as found in the experiment. Since the RKKY interaction is mediated by the conduction electrons, the electron focusing effect could explain the strong directional behavior of the magnetic interaction.

Appendix

This thesis deals with silver (Ag), iron (Fe) and cobalt (Co) impurities buried below a Cu(100) surface and their investigation by low temperature STM. The impurities act as local scattering centers inducing a spatial oscillation in the local density of states (LDOS) with wave vector and amplitude depending on the crystal direction. The interference pattern at the surface is analyzed as function of applied bias voltage. This kind of analysis provides information about the band structure of the host material and the scattering behavior of the atoms, which is for magnetic Fe and Co impurities strongly influenced by the Kondo effect. While the latter one is the most important feature around the Fermi energy, the dispersion of the band structure has to be considered for broader energy intervals.

In the first part of this appendix the energy dependence of the band structure of copper and how this influences the signature of impurities embedded below the Cu(100) surface will be discussed (see Sect. A.1). The second part provides detailed information to the Kondo physics of single sub-surface magnetic Fe and Co atoms. In Sect. A.2 the influence of different tips on the Kondo signature is analyzed. Finally, in Sect. A.3, the dependence of the signature on the tip-surface distance for typical tunneling resistances is experimentally investigated.

A.1 Iso-Energy Surface for Different Energies

A qualitative idea of the influence of the dispersion on the electron focusing effect can be achieved by the iso-energy surface of copper. In Fig. A.1 cross sections for five different energies in the range of $[-0.3\ldots0.3\,\text{eV}]$ are shown. Cross sections of the iso-energy surface along the [010] direction and the [100] direction (Fig. A.1a, b) reveals only a minor change in the considered energy interval. Similar results are obtained for cross sections along the [011] and the [100] direction (Fig. A.1c, d). Neck-like regions are observed for all energies. As a result, the resulting electron propagation and hence the spatial LDOS oscillation may be considered as nearly energy independent within an energy interval of $[-0.1\ldots0.1\,\text{eV}]$. For higher energies the dispersion of the electrons has to be taken into account. This theoretical consideration is in good agreement with the experimental findings for non-magnetic silver impurities (see Sect. 5.2).

H. Prüser, *Scanning Tunneling Spectroscopy of Magnetic Bulk Impurities*,
Springer Theses, DOI: 10.1007/978-3-319-06385-0

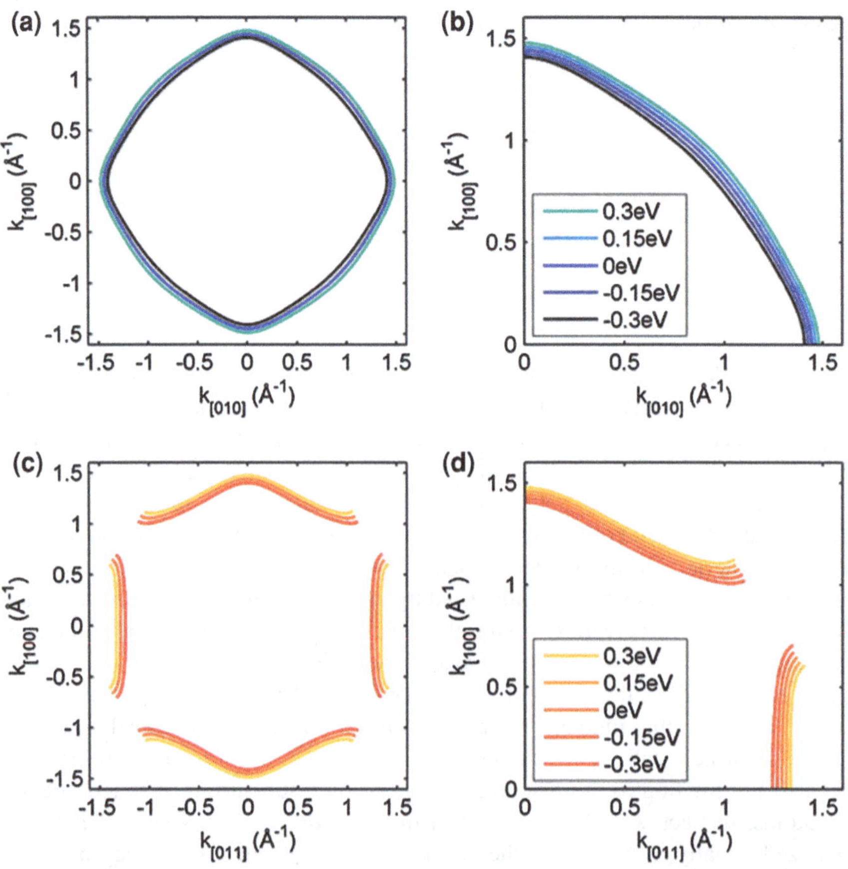

Fig. A.1 Cross sections of the iso-energy surface along different directions as function of energy. **a** Cross section of the iso-energy surface with the [010] direction for an energy range of [−0.3 eV. . .0.3 eV]. **b** A close-up view of the cross sections. **c** Cross section of the iso-energy surface with the [011] direction. **d** A close-up view of the cross sections

A.2 Kondo Signature for Different Tips

In this section the influence of the tip on the Kondo signature for Fe and Co impurities will be discussed. According to the theory (see Sect. 2.2.4) the differential conductance is a convolution of the spectral function of the itinerant electron system—the local density of states—of both tip and sample. For the interpretation of the measured dI/dV signals the tip properties are considered to be energy independent (constant). This simplification is often not fulfilled or has to be considered with caution in the experiment. The shape as well as the chemistry of the apex of the tip can hardly be controlled and may changes from tip to tip or even during the measurement. For this purpose a reference spectrum above the bare

Cu(100) surface far away from the impurity is measured and subtracted. Although this method is a simple approach to deconvolute the differential conductance it is very successful. To illustrate this in Fig. A.2 single $\Delta dI/dV$ curves measured at the center of a fifth monolayer standing-wave interference pattern for three different tips are depicted. The differential conductance directly above the impurity (blue triangles) and above the free surface (green triangles) significantly changes for the three different tips (see Fig. A.2a–c).

In the background subtracted spectra, shown in Fig. A.2d–f, no significant changes for the different tips are observed. Moreover, the extracted microscopic parameters (e.g. the line-shape parameter ϕ, the resonance width Γ, the resonance position ε_k, see Table A.1) are nearly the same for all tips. This shows that the simple background subtraction is able to remove tip-related features.

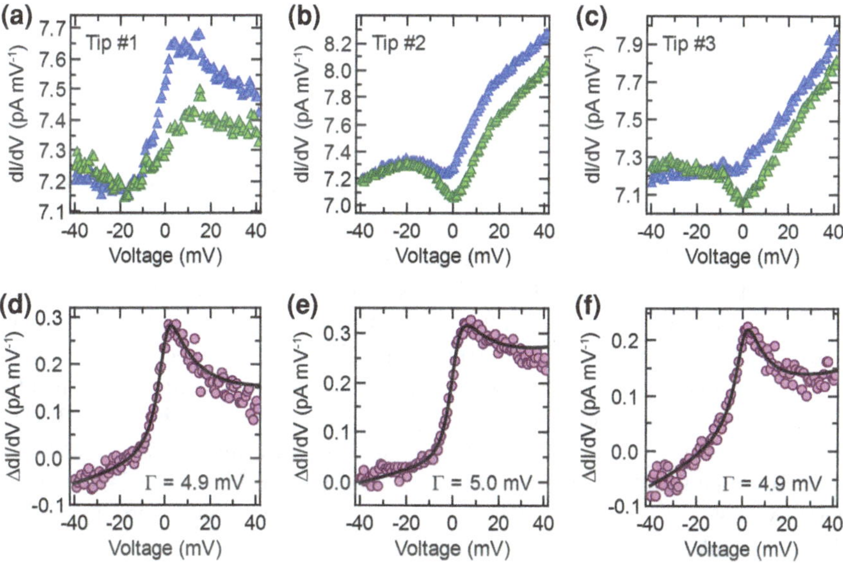

Fig. A.2 Kondo signature in the STM differential conductance for different tips. **a–c** A pair of dI/dV spectra taken with the STM tip held over the center (*blue triangles*) of the interference pattern of a fifth monolayer Fe impurity and over the nearby bare Cu(100) surface (*green triangles*) for different tips. **d–f** Difference of the differential conductance $\Delta dI/dV = dI/dV - dI_0/dV$ (*purple dots*) above the impurity (dI/dV) and above the free surface (dI_0/dV). The solid black curve shows a fit to the $\Delta dI/dV$ data described in Eq. 4.8. Although the spectra of the free surface looks quite different for the different tips the background subtracted signal nearly gives the same spectra for all measurements. Spectroscopic measurements were performed using a lock-in technique with a modulation of the sample voltage of $V_m = 1.3\,\text{mV}\ldots 1.7\,\text{mV}$—at a frequency of 2,190 Hz while the tip height is adjusted to give a stabilizing current of 0.3 nA at $-40\,\text{mV}$

Table A.1 Summary of the fit parameter obtained for different tips

Tip	$\phi\,(\pi)$	Γ (mV)	ε_k (mV)
#1	0.8 ± 0.1	4.9 ± 1.1	0.1 ± 0.8
#2	0.6 ± 0.2	5.0 ± 0.7	0.7 ± 0.8
#3	0.8 ± 0.2	4.9 ± 1.1	0.5 ± 1.4

The corresponding spectroscopic data sets and fits are depicted in Fig. A.2

Table A.2 Summary of the fit parameter obtained for different tip-sample distances

Set point current (nA)	$\phi\,(\pi)$	Γ (mV)	ε_k (mV)
0.05	1.6 ± 0.2	4.6 ± 1.7	-1.5 ± 1.5
0.10	1.8 ± 0.3	4.9 ± 1.4	1.3 ± 1.4
0.25	1.7 ± 0.2	4.7 ± 1.2	0.8 ± 1.2
0.50	1.6 ± 0.2	4.7 ± 1.4	-0.3 ± 1.2
1.00	1.7 ± 0.4	5.0 ± 3.3	0.3 ± 3.0

The corresponding spectroscopic data sets and fits are depicted in Fig. A.3b–f. The results do not depend on the set point current and hence not on the tip-sample distance

A.3 Kondo Signature as Function of Tip-Sample Distances

A detailed analysis of the background subtracted differential conductance shows a clear dependence of the line shape of the Kondo signature as function of impurity depth and the lateral distance. In the experiment an oscillatory behavior is found (see Sect. 4.7). In contrast, a dependence on the tip-surface distance is not observed for typical tunneling resistances used in this work. One example illustrating that the signature does not depend on the tip-sample distance is depicted in Fig. A.3. Here background subtracted $\Delta dI/dV$ spectra are shown, measured at the center of the interference pattern of a third monolayer Fe impurity (see Fig. A.3a). While the set point voltage is fixed to $V_{SP} = -31\,\text{mV}$, the set point currents differs ($I_{SP} = 0.05\,\text{nA}\ldots1.0\,\text{nA}$, see Fig. A.3b–f).

For all set point currents a negative peak around zero bias voltage is found. Assuming an exponential decay of the tunneling current

$$I \propto \exp\left(-2\sqrt{\frac{2m_e\Phi}{\hbar^2}}z\right) \tag{A.1}$$

with tip-sample distance z and a mean work function of $\Phi = 4.2\,\text{eV}$ (see Sect. 2.2.6), a change of the tunneling current from $I = 0.05\,\text{nA}$–$1.00\,\text{nA}$ correspond to a change of the tip-sample distance of $\Delta z = 1.4\,\text{Å}$.

Despite some small variations all spectra for the different distances reveal the same Kondo signature. Applying the phenomenological fit formula, given in Eq. 4.8, to the data yields nearly the same microscopic parameters (e.g. the line-shape

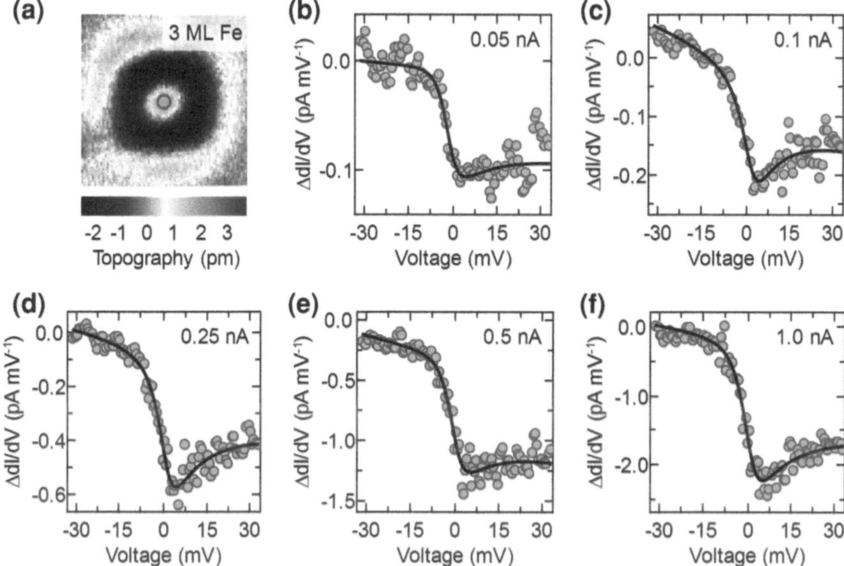

Fig. A.3 Kondo signature as function of tip-sample distances. **a** Topography of a third monolayer Fe impurity (1.5 × 1.5 nm, −31 mV, 0.3 nA). **b–f** Single $\Delta dI/dV$ spectra (*green curve*) measured at the center of the interference pattern (marked by the *green dot* in **a**) for different set point currents of 0.05 nA, 0.1 nA, 0.25 nA, 0.5 nA and 1.0 nA (**b–f** respectively). All $\Delta dI/dV$ data sets show the same signature. The spectroscopic measurements were performed using a lock-in technique with a modulation of the sample voltage of $V_m = 0.7$ mV at a frequency of 2,556 Hz. The set point voltage $V_{SP} = -31$ mV is constant for all data sets. The *solid black curve* shows a fit to the $\Delta dI/dV$ data described in Eq. 4.8.

parameter ϕ, the resonance width Γ, the resonance position ε_k, see Table A.2). Considering not only this but also other data sets show that the observed spectroscopic signature do not depend on typical tunneling resistances in the order of $0.01\,\mathrm{G}\Omega\ldots100\,\mathrm{G}\Omega$ and hence typical tip-sample distances used in this work.

Zeitfracht Medien GmbH
Ferdinand-Jühlke-Straße 7
99095 Erfurt, Deutschland
produktsicherheit@kolibri360.de